Algebra II Workbook

FOR DUMMIES®

A Wiley Brand

2nd Edition

by Mary Jane Sterling

FOR DUMMIES®
A Wiley Brand

Algebra II Workbook For Dummies, 2nd Edition

Published by: **John Wiley & Sons, Inc.**, 111 River Street, Hoboken, NJ 07030-5774, www.wiley.com

Copyright © 2014 by John Wiley & Sons, Inc., Hoboken, New Jersey

Published simultaneously in Canada

For general information on our other products and services, please contact our Customer Care Department within the U.S. at 877-762-2974, outside the U.S. at 317-572-3993, or fax 317-572-4002. For technical support, please visit www.wiley.com/techsupport.

Wiley publishes in a variety of print and electronic formats and by print-on-demand. Some material included with standard print versions of this book may not be included in e-books or in print-on-demand. If this book refers to media such as a CD or DVD that is not included in the version you purchased, you may download this material at http://booksupport.wiley.com. For more information about Wiley products, visit www.wiley.com.

Library of Congress Control Number: 2013956855

ISBN 978-1-118-86703-7 (pbk); ISBN 978-1-118-86708-2 (ebk); ISBN 978-1-118-86685-6 (ebk)

Manufactured in the United States of America

10 9 8 7 6 5 4 3 2 1

Contents at a Glance

Table of Contents

Introduction

Here you are, pencil in hand, ready to take on the challenges of working on Algebra II problems. How did you get here? Are you taking an Algebra II class and just not getting enough homework assigned? Or have you found a few gaps in the instruction and want to fill them in before you end up with a flood of questions? Maybe you've been away from algebra for a while and you want a review. Or perhaps you're getting ready to tackle another mathematics course, such as Calculus. If you're looking for some good-natured, clear explanations on how to do some standard and challenging algebra problems, then you've come to the right place.

I hope you can find everything you need in this book to practice the concepts of Algebra II. You'll find some basic (to get you in the mood) and advanced algebra topics. But not all the basics are here — that's where Algebra I comes in. The topics that aren't here are referenced for your investigation or further study.

Calculus and other more-advanced math drive Algebra II. Algebra is the passport to studying calculus and trigonometry and number theory and geometry and all sorts of good mathematics and science. Algebra is basic, and the algebra here can help you grow in your skills and knowledge.

About This Book

You don't have to do the problems in this book in the order in which they're presented. You can go to the topics you want or need and refer back to earlier problems if necessary. You can jump back and forth and up and down, if so inclined (but please, not on the furniture). The organization allows you to move freely about and find what you need.

Use this book as a review or to supplement your study of Algebra II. Each section has a short explanation and an example or two — enough information to allow you to do the problems.

If you want more background or historical information on a topic, you can refer to the companion book, *Algebra II For Dummies,* where I go into more depth on what's involved with each type of problem. (If you need more-basic information, you can try *Algebra I For Dummies* and *Algebra I Workbook For Dummies*). In this workbook, I get to the point quickly but with enough detail to see you through. The answers to the problems, at the end of each chapter, provide even more step-by-step instruction.

Foolish Assumptions

You're interested in doing algebra problems. Is that a foolish thing for me to assume? No! Of course, you're interested and excited and, perhaps, just a slight bit tentative. No need to worry. In this book, I assume that you have a decent background in the basics of algebra and want to investigate further. If so, this is the place to be. I take those basic concepts and expand your horizons in the world of algebra.

Are you a bit rusty with your algebra skills? Then the worked-out solutions in this book will act as refreshers as you investigate the different topics. You may be preparing for a more advanced mathematics course such as trigonometry or calculus. Again, the material in this book will be helpful.

Or maybe it's just my first assumption that fits your situation: You're interested in doing algebra and couldn't pass up doing the problems in this book!

Icons Used in This Book

Throughout this book, I highlight some of the most important information with icons. Here's what the icons mean:

 You can read the word *rules* as a noun or a verb. Sometimes it's hard to differentiate. But usually, in this book, *rules* is a noun. This icon marks a formula or theorem or law from algebra that pertains to the subject at hand. The rule applies at that moment and at any moment in algebra.

 You see this icon when I present an example problem whose solution I walk you through step by step. You get a problem and a detailed answer.

 This icon refers back to information that I think you may already know. It needs to be pointed out or repeated so that the current explanation makes sense.

 Tips show you a quick and easy way to do a problem. Try these tricks as you're solving problems.

 There are always things that are tricky or confusing or problems that just ask for an error to happen. This icon is there to alert you, hoping to help you avoid a mathematical pitfall.

Beyond the Book

In addition to the material in the print or e-book you're reading right now, this product also comes with some access-anywhere goodies on the web. No matter how diligent you are about reading through this material, you'll likely come across a few questions where you don't have a clue. Check out the free Cheat Sheet at `www.dummies.com/cheatsheet/algebraiiwb` for helpful information, all provided in a concise, quick-access format.

Finally, you can find some articles online that tie together and offer new insights to the material you find in this book. Go to `www.dummies.com/extras/algebraiiwb` for these informative articles.

Where to Go from Here

You may become intrigued with a particular topic or particular type of problem. Where do you find more problems like those found in a section? Where do you find the historical background of a favorite algebra process? There are many resources out there, including a couple that I wrote:

- ✔ Do you like the applications? Try *Math Word Problems For Dummies.*
- ✔ Are you more interested in the business-type uses of algebra? Take a look at *Business Math For Dummies.*

If you're ready for another area of mathematics, look for a couple more of my titles: *Trigonometry For Dummies* and *Linear Algebra For Dummies.*

Part I
Getting Started with Algebra II

In this part . . .

- ✔ Find order in the order of operations and relate algebraic properties to processes used when solving equations.

- ✔ Solve linear equations and inequalities and rewrite absolute value equations before solving.

- ✔ Take on radical equations, rational equations, and fractional exponents.

- ✔ Use one or more factorization methods to ready equations for the multiplication property of zero.

- ✔ Solve equations with the quadratic formula or complete the square.

- ✔ Graph basic curves using intercepts and properties of functions.

Chapter 1

Going Beyond Beginning Algebra

In This Chapter

▶ Applying order of operations and algebraic properties

▶ Using FOIL and other products

▶ Solving linear and absolute value equations

▶ Dealing with inequalities

he nice thing about the rules in algebra is that they apply no matter what level of mathematics or what area of math you're studying. Everyone follows the same rules, so you find a nice consistency and orderliness. In this chapter, I discuss and use the basic rules to prepare you for the topics that show up in Algebra II.

Good Citizenship: Following the Order of Operations and Other Properties

The *order of operations* in mathematics deals with what comes first (much like the chicken and the egg). When faced with multiple operations, this *order* tells you the proper course of action.

The *order of operations* states that you use the following sequence when simplifying algebraic expressions:

1. **Raise to powers or find roots.**

2. **Multiply or divide.**

3. **Add or subtract.**

Special groupings can override the normal order of operations. For instance, a^{b+c} asks you to add $b + c$ before raising a to the power, which is a sum. If groupings are a part of the expression, first perform whatever's in the grouping symbol. The most common grouping symbols are parentheses, ( ); brackets, [ ]; braces, { }; fraction bars, —; absolute value bars, | |; and radical signs, $\sqrt{}$.

If you find more than one operation from the same level, move from left to right performing those operations.

The commutative, associative, and distributive properties allow you to rewrite expressions and not change their value. So, what do these properties say? Great question! And here are the answers:

- ✔ **Commutative property of addition and multiplication:** $a + b = b + a$, and $a \cdot b = b \cdot a$; the order doesn't matter.

    Rewrite subtraction problems as addition problems so you can use the commutative (and associative) property. In other words, think of $a - b$ as $a + (-b)$.

- ✔ **Associative property of addition and multiplication:** $a + (b + c) = (a + b) + c$, and $a(b \cdot c) = (a \cdot b)c$; the order is the same, but the grouping changes.

- ✔ **Distributive property of multiplication over addition (or subtraction):** $a(b + c) = a \cdot b + a \cdot c$, and $a(b - c) = a \cdot b - a \cdot c$.

The *multiplication property of zero* states that if the product of $a \cdot b = 0$, then either a or b (or both) must be equal to 0.

Q. Use the order of operations and other properties to simplify the expression $\dfrac{12\left(\left(\frac{5}{6} + \frac{3}{4}\right) - \frac{1}{6}\right)}{5\sqrt{2(3)^2 + 7}}$.

A. $\dfrac{17}{25}$. The big fraction bar is a grouping symbol, so you deal with the numerator and denominator separately. Use the commutative and associative properties to rearrange the fractions in the numerator; square the 3 under the radical in the denominator. Next, in the numerator, combine the fractions that have a common denominator; below the fraction bar, multiply the two numbers under the radical. Reduce the first fraction in the numerator; add the numbers under the radical. Distribute the 12 over the two fractions; take the square root in the denominator. Simplify the numerator and denominator.

Here's what the process looks like:

$$\frac{12\left(\left(\frac{5}{6} - \frac{1}{6}\right) + \frac{3}{4}\right)}{5\sqrt{2(9) + 7}} = \frac{12\left(\left(\frac{4}{6}\right) + \frac{3}{4}\right)}{5\sqrt{18 + 7}} = \frac{12\left(\frac{2}{3} + \frac{3}{4}\right)}{5\sqrt{25}} = \frac{8 + 9}{5(5)} = \frac{17}{25}$$

1. Simplify: $-3[4 - 2(3^2)]$

2. Simplify: $4^3(3^2 + 11)(7^4 - 7^4)(10^{10})$

3. Simplify: $153 + 187 - 153 + 270 - 471 - 270 + 471$

4. Simplify: $(6^2 + 4)\sqrt{5^2 - 3^2}$

5. Simplify: $\dfrac{\sqrt[3]{4(3^2 + 7)}}{\sqrt{121} - \sqrt{49}}$

6. Simplify: $7 + 3\left|2(8) - 5^2\right|$

(For info on absolute value, see the upcoming section, "Dealing with Linear Absolute Value Equations.")

Specializing in Products and FOIL

Multiplying algebraic expressions together can be dandy and nice or downright gruesome. Taking advantage of patterns and processes makes the multiplication quicker, easier, and more accurate.

When multiplying two binomials together, you have to multiply the two terms in the first binomial times the two terms in the second binomial — you're actually *distributing* the first terms over the second. The FOIL acronym describes a way of multiplying those terms in an organized fashion, saving space and time. *FOIL* refers to multiplying the two *First* terms together, then the two *Outer* terms, then the two *Inner* terms, and finally the two *Last* terms. The *Outer* and *Inner* terms usually combine. Then you add the products together by combining like terms. So, if you have $(ax + b)(cx + d)$, you can do the multiplication of the terms, or FOIL, like so:

Terms	*Product*
First	$ax(cx)$
Outer	$ax(d)$
Inner	$b(cx)$
Last	$b(d)$

$$= acx^2 + adx + bcx + bd$$
$$= acx^2 + (ad + bc)x + bd$$

The following examples show some multiplication patterns to use when multiplying binomials (expressions with two terms).

Q. Find the square of the binomial: $(2x + 3)^2$

A. **$4x^2 + 12x + 9$.** When squaring a binomial, you square both terms and put twice the product of the two original terms between the squares: $(a + b)^2 = a^2 + 2ab + b^2$. So, $(2x + 3)^2 = (2x)^2 + 2(2x)(3) + 3^2 = 4x^2 + 12x + 9$.

Q. Multiply the two binomials together using FOIL: $(3x - 5)(4x + 7)$

A. **$12x^2 + x - 35$.** Find the products: *First*, $3x(4x) = 12x^2$, plus *Outer*, $3x(7) = 21x$, plus *Inner*, $(-5)(4x) = -20x$, plus *Last*, $(-5)(7) = -35$. So, the product of $(3x - 5)(4x + 7)$ is $(3x)(4x) + (3x)(7) + (-5)(4x) + (-5)(7) = 12x^2 + 21x - 20x - 35 = 12x^2 + x - 35$.

Q. Find the product of the binomial and the trinomial: $(2x + 7)(3x^2 + x - 5)$

A. **$6x^3 + 23x^2 - 3x - 35$.** Distribute the $2x$ over the terms in the trinomial, and then distribute the 7 over the same terms. Combine like terms to simplify. The product of $(2x + 7)(3x^2 + x - 5)$ is $2x(3x^2 + x - 5) + 7(3x^2 + x - 5) = 6x^3 + 2x^2 - 10x + 21x^2 + 7x - 35 = 6x^3 + 23x^2 - 3x - 35$.

7. Square the binomial: $(4x - 5)^2$

8. Multiply: $(5y - 6)(5y + 6)$

9. Multiply: $(8z - 3)(2z + 5)$

10. Multiply: $(2x - 5)(4x^2 + 10x + 25)$

Variables on the Side: Solving Linear Equations

A linear equation has the general format $ax + b = c$, where x is the variable and a, b, and c are constants. When you solve a linear equation, you're looking for the value that x takes on to make the linear equation a true statement. The general game plan for solving linear equations is to isolate the term with the variable on one side of the equation and then multiply or divide to find the solution.

Q. Solve for x in the equation $\frac{3(x+7)-5}{4} = 2x + 9$.

A. $x = -4$. First, multiply each side by 4 to get rid of the fraction. Then distribute the 3 over the terms in the parentheses. Combine the like terms on the left. Next, you want all variable terms on one side of the equation, so subtract $8x$ and 16 from each side. Finally, divide each side by –5.

$$\frac{4}{1} \cdot \frac{3(x+7)-5}{4} = 4(2x+9)$$
$$3(x+7)-5 = 8x+36$$
$$3x+21-5 = 8x+36$$
$$3x+16 = 8x+36$$
$$-5x = 20$$
$$x = -4$$

---

11. Solve for x: $4x - 5 = 3(3x + 10)$

12. Solve for x: $5[9 - x(x + 8)] = 5x(1 - x)$

---

13. Solve for x: $\frac{x}{4} + \frac{x-6}{5} = \frac{x+2}{3}$

14. Solve for x: $(x - 1) + 2(x - 2) + 3(x - 3) + 4(x - 4) + 5(x - 5) = 0$

Dealing with Linear Absolute Value Equations

The absolute value of a number is the number's distance from 0. The formal definition of *absolute value* is

$$|a| = \begin{cases} a, a \geq 0 \\ -a, a < 0 \end{cases}$$

In other words, the absolute value of a number is exactly that number unless the number is negative; when the number is negative, its absolute value is the opposite, or a positive. The absolute value of a number, then, is the number's value without a sign; it's never negative.

When solving linear absolute value equations, you have two possibilities: one that the quantity *inside* the absolute value bars is positive, and the other that it's negative. Because you have to consider both situations, you usually get two different answers when solving absolute value equations, one from each scenario. The two answers come from setting the quantity inside the absolute value bars first equal to a positive value and then equal to a negative value.

Before setting the quantity equal to the positive and negative values, first isolate the absolute value term on one side of the equation by adding or subtracting the other terms (if you have any) from each side of the equation.

TIP

If you find more than one absolute value expression in your problem, you have to get down and dirty — consider all the possibilities. A value inside absolute value bars can be either positive or negative, so look at all the different combinations: Both values within the bars are positive, or the first is positive and the second is negative, or the first is negative and the second positive, or both are negative. Whew!

EXAMPLE

Q. Solve for x in $|3x + 7| = 11$.

A. $x = \frac{4}{3}, -6, -6.$ First rewrite the absolute value equation as two separate linear equations. In the first equation, assume that the $3x + 7$ is positive and set it equal to 11. In the second equation, also equal to 11, assume that the $(3x + 7)$ is negative. For that one, negate (multiply by -1) the whole binomial, and then solve the equation.

$$3x + 7 = 11$$
$$3x = 4$$
$$x = \frac{4}{3}$$

$$-(3x + 7) = 11$$
$$-3x - 7 = 11$$
$$-3x = 18$$
$$x = -6$$

---

15. Solve for x in $|2x + 1| = 83$.

16. Solve for x in $5|4 - 3x| + 6 = 11$.

17. Solve for x in $|x + 3| + |2x - 1| = 6$.

18. Solve for x in $3|4x - 5| + 10 = 7$.

Greater Math Skills: Equalizing Linear Inequalities

A linear inequality resembles a linear equation — except for the relationship between the terms. The basic form for a linear inequality is $ax + b > c$. When $\geq$ or $<$ or $\leq$ are in the statement, the methods used to solve the inequality stay the same. When the extra bar appears under the inequality symbol, it means "or equal to," so you read $\leq$ as "is less than or equal to."

The main time to watch out when solving inequalities is when you multiply or divide each side of the inequality by a negative number. When you do that — and yes, you're allowed — you have to reverse the sense or the relationship. The inequality $>$ becomes $<$, and vice versa.

When solving *absolute value inequalities* (see the preceding section for more on absolute values), you first drop the absolute value bars. Then you apply one of two separate rules for absolute value inequalities, depending on which way the inequality symbol faces:

✔ Solve $|ax + b| > c$ using the two inequalities $ax + b > c$ and $ax + b < -c$.

✔ Solve $|ax + b| < c$ using the single compound inequality $-c < ax + b < c$.

Two ways of writing your answers are inequality notation and interval notation:

✔ **Inequality notation:** This notation is just what it says it is: If your answer is all x's greater than or equal to 3, you write $x \geq 3$. To say that the answer is all x values between -5 and $+5$, including the -5 but not the positive 5, you write $-5 \leq x < 5$.

✔ **Interval notation:** Some mathematicians prefer interval notation because it's so short and sweet. You simply list the starting and stopping points of the numbers you want to use. When you see this notation, you just have to recognize that you're discussing intervals of numbers (and not, for instance, the coordinates of a point). The rule is that you use a bracket, [or], when you want to include the number, and use a parenthesis, (or), when you don't want to include the number. You always use a parenthesis with ∞ or $-\infty$. For example, to write $x \geq 3$ in interval notation, you use $[3, \infty)$. Writing $-5 \leq x < 5$ in interval notation, you have $[-5, 5)$.

Q. Solve the inequality $8x - 15 \geq 10x + 7$. Write the answer in both inequality and interval notation.

A. $x \leq -11$ **and** $(-\infty, -11]$. First subtract $10x$ from each side; then add 15 to each side. This step moves the variable terms to the left and the constants to the right: $-2x \geq 22$. Now divide each side by -2. Because you're dividing by a negative number, you need to reverse the inequality sign: $x \leq -11$. That's the answer in inequality notation. The solution is that x can be any number either equal to or smaller than -11. In interval notation, you write this as $(-\infty, -11]$.

Q. Solve the inequality $|5 - 6x| < 7$. Write the answer in both inequality and interval notation.

A. $-\frac{1}{3} < x < 2$ **and** $\left(-\frac{1}{3}, 2\right)$. Rewrite the absolute value inequality as the inequality $-7 < 5 - 6x < 7$. Subtract 5 from each of the three sections of the inequality to put the variable term by itself: $-12 < -6x < 2$. Now divide each section by -6, reversing the inequality symbols. Then, after you've gone to all the trouble of reversing the inequalities, rewrite the statement again with the smaller number on the left to correspond to numbers on the number line. This step requires reversing the inequalities again. Here are the details:

$$-12 < -6x < 2$$
$$2 > x > -\frac{1}{3}$$
$$-\frac{1}{3} < x < 2$$

In interval notation, the answer is $\left(-\frac{1}{3}, 2\right)$.

This answer looks very much like the coordinates of a point. In instances like this, be very clear about what you're trying to convey with the interval notation.

---

19. Solve the inequality $5x + 7 \leq 22$.

20. Solve the inequality $8(y - 4) \geq 5(3y - 12)$.

21. Solve the inequality $|3x + 7| > 4$.

22. Solve the inequality $4|5 - 2x| + 3 < 7$.

Answers to Problems on Going Beyond Beginning Algebra

This section provides the answers (in bold) to the practice problems in this chapter.

1 Simplify: $-3[4 - 2(3^2)]$. The answer is **42.**

$-3[4 - 2(9)] = -3[4 - 18] = -3[-14] = 42$

2 Simplify: $4^3(3^2 + 11)(7^4 - 7^4)(10^{10})$. The answer is **0.**

The third factor is 0. This makes the whole product equal to 0. Remember, the *multiplication property of zero* says that if any factor in a product is equal to 0, then the entire product is equal to 0.

3 Simplify: $153 + 187 - 153 + 270 - 471 - 270 + 471$. The answer is **187.**

Use the associative and commutative properties to write the numbers and their opposites together:

$187 + 153 - 153 + 270 - 270 - 471 + 471 = 187 + 0 + 0 + 0 = 187$

4 Simplify: $\left(6^2 + 4\right)\sqrt{5^2 - 3^2}$. The answer is **160.**

$$(36 + 4)\sqrt{25 - 9} = (40)\sqrt{16} = (40)(4) = 160$$

5 Simplify: $\dfrac{\sqrt[3]{4(3^2 + 7)}}{\sqrt{121} - \sqrt{49}}$. The answer is **1.**

$$\frac{\sqrt[3]{4(9 + 7)}}{\sqrt{121} - \sqrt{49}} = \frac{\sqrt[3]{4(16)}}{11 - 7} = \frac{\sqrt[3]{64}}{4} = \frac{4}{4} = 1$$

6 Simplify: $7 + 3|2(8) - 5^2|$. The answer is **34.**

$7 + 3|16 - 25| = 7 + 3|-9| = 7 + 3(9) = 7 + 27 = 34$

7 Square the binomial: $(4x - 5)^2$. The answer is $\mathbf{16x^2 - 40x + 25.}$

When squaring a binomial, the two terms are each squared, and the term between them is twice the product of the original terms.

A common error in squaring binomials is to forget the middle term and just use the squares of the two terms in the binomial. If you tend to forget the middle term, you can avoid the error and get the correct answer through FOIL — $(4x - 5)(4x - 5) = 16x^2 - 20x - 20x + 25 = 16x^2 - 40x + 25$.

8 Multiply: $(5y - 6)(5y + 6)$. The answer is $\mathbf{25y^2 - 36.}$

The product of two binomials that contain the sum and difference of the same two terms results in a binomial that's the difference between the squares of the terms.

9 Multiply: $(8z - 3)(2z + 5)$. The answer is $\mathbf{16z^2 + 34z - 15.}$

Using FOIL, the first term in the answer is the product of the first two terms: $(8z)(2z)$. The middle term is the sum of the products of the *Outer* and *Inner* terms: $(8z)(5)$ and $(-3)(2z)$. The final term is the product of the two last terms: $(-3)(5)$.

10 Multiply: $(2x - 5)(4x^2 + 10x + 25)$. The answer is $\mathbf{8x^3 - 125}$.

Distribute the first term in the binomial $(2x)$ over the terms in the trinomial, and then distribute the second term in the binomial (-5) over the terms in the trinomial. After that, combine like terms:

$$= 2x(4x^2 + 10x + 25) - 5(4x^2 + 10x + 25)$$
$$= 8x^3 + 20x^2 + 50x - 20x^2 - 50x - 125$$
$$= 8x^3 + 20x^2 - 20x^2 + 50x - 50x - 125$$
$$= 8x^3 - 125$$

11 Solve for x: $4x - 5 = 3(3x + 10)$. The answer is $\mathbf{x = -7}$.

Distribute the terms on the right to get $4x - 5 = 9x + 30$. Subtract $9x$ and add 5 to each side, which gives you $-5x = 35$. Then divide each side by -5: $x = -7$.

12 Solve for x: $5[9 - x(x + 8)] = 5x(1 - x)$. The answer is $\mathbf{x = 1}$.

First distribute the x in the bracket on the left and the $5x$ on the right. Then you can distribute the 5 outside the bracket on the left to see what the individual terms are. The squared terms disappear when you add $5x^2$ to each side of the equation. Solve for x:

$$5\left[9 - x^2 - 8x\right] = 5x - 5x^2$$
$$45 - 5x^2 - 40x = 5x - 5x^2$$
$$45 - 40x = 5x$$
$$45 = 45x$$
$$1 = x$$

13 Solve for x: $\dfrac{x}{4} + \dfrac{x-6}{5} = \dfrac{x+2}{3}$. The answer is $\mathbf{x = 16}$.

First, multiply each fraction by 60, the least common multiple. Then distribute and simplify:

$$\overset{15}{\cancel{60}}\left(\frac{x}{\cancel{4}_1}\right) + \overset{12}{\cancel{60}}\left(\frac{x-6}{\cancel{5}_1}\right) = \overset{20}{\cancel{60}}\left(\frac{x+2}{\cancel{3}_1}\right)$$
$$15x + 12(x - 6) = 20(x + 2)$$
$$15x + 12x - 72 = 20x + 40$$
$$7x = 112$$
$$x = 16$$

14 Solve for x: $(x - 1) + 2(x - 2) + 3(x - 3) + 4(x - 4) + 5(x - 5) = 0$. The answer is $\mathbf{x = \dfrac{11}{3}}$.

Distribute over the terms. Combine like terms and solve for x.

$$x - 1 + 2x - 4 + 3x - 9 + 4x - 16 + 5x - 25 = 0$$
$$15x - 55 = 0$$
$$15x = 55$$
$$x = \frac{11}{3}$$

15 Solve for x in $|2x + 1| = 83$. The answer is $\mathbf{x = 41, -42}$.

First, let the value in the absolute value be positive, and solve $2x + 1 = 83$. You get $2x = 82$, or $x = 41$. Next, let the value in the absolute value be negative, and solve $-(2x + 1) = 83$. (If you multiply each side by -1, you don't have to distribute the negative sign over two terms. Instead, you have $2x + 1 = -83$.) Solving this, you get $2x = -84$, or $x = -42$.

16 Solve for x in $5|4 - 3x| + 6 = 11$. The answer is $x = 1, \frac{5}{3}$.

First subtract 6 from each side, and then divide by 5. (You can't apply the rule for changing an absolute value equation into linear equations unless the absolute value is isolated on one side of the equation.) Now you have $|4 - 3x| = 1$. Letting the expression inside the absolute value be positive, you have $4 - 3x = 1$, which means $-3x = -3$ and $x = 1$. Now, if the expression inside the absolute value is negative, you make the expression positive by negating the whole thing: $-(4 - 3x) = 1$ gives you $4 - 3x = -1$, which means $-3x = -5$ and $x = \frac{5}{3}$.

17 Solve for x in $|x + 3| + |2x - 1| = 6$. The answer is $x = \frac{4}{3}, -2$.

You need four different equations to solve this problem. Consider that both absolute values may be positive; then that the first is positive and the second, negative; then that the first is negative and the second, positive; and last, that both are negative.

Unfortunately, not every equation gives you an answer that really works. Perhaps no value of x can make the first absolute value negative and the second positive. An extraneous solution to an equation or inequality is a false or incorrect solution. It occurs when you change the original format of the equation to a form that's more easily solved. The extraneous solution may be a solution to the new form, but it doesn't work in the original. After you change the format, simply solve each equation produced, and then check each answer in the original equation.

Here are the situations, one at a time:

- **Positive/Positive:** $(x + 3) + (2x - 1) = 6$; $x + 3 + 2x - 1 = 6$; $3x + 2 = 6$; $3x = 4$; $x = \frac{4}{3}$. When you put this answer back into the original equation, it works — you get a true statement.

- **Positive/Negative:** $(x + 3) + -(2x - 1) = 6$; $x + 3 - 2x + 1 = 6$; $-x + 4 = 6$; $-x = 2$; $x = -2$. Putting this value back into the original equation, you get a true statement, so this answer is valid, too.

- **Negative/Positive:** $-(x + 3) + (2x - 1) = 6$; $-x - 3 + 2x - 1 = 6$; $x - 4 = 6$; $x = 10$. When you put this back into the original equation, you get $13 + 19 = 6$. That isn't true, so this solution is extraneous. It doesn't work.

- **Negative/Negative:** $-(x + 3) + -(2x - 1) = 6$; $-x - 3 - 2x + 1 = 6$; $-3x - 2 = 6$; $-3x = 8$; $x = -\frac{8}{3}$.

  You get another extraneous solution. When you put this value for x back into the original equation, you get

$$\left| -\frac{8}{3} + 3 \right| + \left| 2\left(-\frac{8}{3}\right) - 1 \right| \stackrel{?}{=} 6$$

$$\left| \frac{1}{3} \right| + \left| -\frac{19}{3} \right| \stackrel{?}{=} 6$$

$$\frac{20}{3} \neq 6$$

  Not so! This solution doesn't work.

18 Solve for x in $3|4x - 5| + 10 = 7$. **No solution.**

This absolute value equation has no solution, because it asks you to find some number whose absolute value is negative. When you subtract 10 from each side and divide each side by 3, you get $|4x - 5| = -1$. That's impossible. The absolute value of any number is either positive or 0; it's never negative.

19 Solve the inequality $5x + 7 \leq 22$. The answer is $x \leq 3$.

First, subtract 7 from each side to get $5x \leq 15$. Then divide each side by 5. In interval notation, write the solution as $(-\infty, 3]$.

20 Solve the inequality $8(y - 4) \geq 5(3y - 12)$. The answer is $y \leq 4$.

Distribute the 8 and 5 over their respective binomials to get $8y - 32 \geq 15y - 60$. Subtract $15y$ from each side and add 32 to each side: $-7y \geq -28$. When you divide each side by -7, you reverse the inequality to get $y \leq 4$. Write this answer as $(-\infty, 4]$ in interval notation.

21 Solve the inequality $|3x + 7| > 4$. The answer is $x > -1$ or $x < -\dfrac{11}{3}$.

Rewrite the absolute value inequality as two separate linear inequalities: $3x + 7 > 4$ and $3x + 7 < -4$.

Solving the first, you get $3x > -3$, or $x > -1$. Solving the second, you get $3x < -11$, or $x < -\dfrac{11}{3}$. In interval notation, you write this solution as $(-1, \infty)$ or $\left(-\infty, -\dfrac{11}{3}\right)$.

22 Solve the inequality $4|5 - 2x| + 3 < 7$. The answer is $2 < x < 3$.

First, subtract 3 from each side and then divide each side by 4 to get $|5 - 2x| < 1$. Then rewrite the absolute value inequality as $-1 < 5 - 2x < 1$. Subtract 5 from each interval: $-6 < -2x < -4$. Dividing each interval by -2 in the next step means reversing the inequality signs: $3 > x > 2$. The numbers should be written from smallest to largest, so switch the numbers and their inequality signs to get $2 < x < 3$. In interval notation, you write this answer as $(2, 3)$.

Chapter 2

Handling Quadratic (and Quadratic-Like) Equations and Inequalities

In This Chapter

▶ Finding solutions with radicals

▶ Solving quadratic equations using factoring and the quadratic formula

▶ Completing the square

▶ Changing equations with higher powers to quadratic form

▶ Dealing with quadratic inequalities using number lines

uadratic equations and inequalities include variables that have powers, or exponents, of 2. The power 2 opens up the possibilities for more solutions than do linear equations (whose variables have powers of 1 — see Chapter 1). For instance, the linear equation $2x + 3 = 5$ has one solution, $x = 1$, but the quadratic equation $2x^2 + 3 = 5$ has two solutions, $x = 1$ and $x = -1$. You can solve quadratic equations through factoring, employing the quadratic formula, completing the square, or using the nifty *square root rule* when possible. Quadratic inequalities, on the other hand, are best solved by looking at intervals on a number line.

Some of the equations in this chapter go beyond the second degree (power or exponent on the variable) — they start out with a degree of 4 or 6 or more, but a little tweaking brings you back to the basic quadratic equation and its many possibilities. (For information on higher-powered equations that you can't change to quadratic form, see Chapter 7, on polynomials.)

Finding Reasonable Solutions with Radicals

When a quadratic equation consists of just a squared term and a constant, you use the *square root rule* to quickly solve the equation.

Solve the equation $ax^2 = k$ by dividing each side of the equation by a and then taking the square root of each side: $ax^2 = k$, $x^2 = \dfrac{k}{a}$, $x = \pm\sqrt{\dfrac{k}{a}}$. Note that you end up with two roots — one positive and one negative.

Q. Solve $9x^2 + 11 = 27$ using the square root rule.

A. $x = \pm\dfrac{4}{3}$. Subtract 11 from each side, and you get $9x^2 = 16$. Then divide each side of the equation by 9, take the square root of each side, and simplify: $x^2 = \dfrac{16}{9}$, $x = \pm\sqrt{\dfrac{16}{9}} = \pm\dfrac{\sqrt{16}}{\sqrt{9}} = \pm\dfrac{4}{3}$. The equation gives you one positive and one negative solution.

Q. Solve $3x^2 = 36$ using the square root rule.

A. $x = \pm 2\sqrt{3}$. Divide each side by 3, and then take the square root of each side: $x^2 = 12$; $x = \pm\sqrt{12}$. You can simplify the radical because of the following property: $\sqrt{a \cdot b} = \sqrt{a} \cdot \sqrt{b}$. Thus, the equation becomes $x = \pm\sqrt{4 \cdot 3} = \pm\sqrt{4}\sqrt{3} = \pm 2\sqrt{3}$.

1. Solve for x: $4x^2 = 25$

2. Solve for x: $3x^2 + 8 = 56$

3. Solve for x: $x^2 = 7$

4. Solve for x: $5x^2 = 200$

UnFOILed Again! Successfully Factoring for Solutions

The quickest, easiest way to solve a quadratic equation is to factor it and set the individual factors equal to 0. Of course, the expression in the equation has to be factorable. If it isn't, you can rely on that old standby, the quadratic formula (see "Resorting to the Quadratic Formula," later in this chapter). In either case, make sure your quadratic equation is in the correct form before you begin.

You can factor a quadratic equation in the form $ax^2 + bx + c = 0$ if you can find two binomials $(dx + e)(fx + g)$ whose product is $ax^2 + bx + c$. Factoring is like working out a puzzle to figure out what the coefficients and constants are.

If $ax^2 + bx + c = (dx + e)(fx + g)$, then $d \cdot f = a$ and $e \cdot g = c$; the sum of $d \cdot g$ and $e \cdot f$ is b. This process is essentially undoing FOIL. (*FOIL* is the acronym for remembering how to multiply the terms in two binomials together: *F*irst, *O*uter, *I*nner, *L*ast — see Chapter 1 for details.)

Knowing whether to use + or – signs in the binomials can really ease the factoring process. Table 2-1 shows you how the order of the + and – signs in the quadratic equation can give you clues about the signs that show up in the factors.

Table 2-1	Signs to Use in the Binomials When Factoring	
Quadratic Equation	Signs in the Binomials	Example
$ax^2 + bx + c$	$(+)(+)$	$x^2 + 5x + 6 = (x + 2)(x + 3)$
$ax^2 - bx + c$	$(-)(-)$	$x^2 - 5x + 6 = (x - 2)(x - 3)$
$ax^2 + bx - c$	$(+)(-)$ or $(-)(+)$	$x^2 + x - 6 = (x + 3)(x - 2)$
$ax^2 - bx - c$	$(+)(-)$ or $(-)(+)$	$x^2 - x - 6 = (x + 2)(x - 3)$

Q. Solve by factoring: $8x^2 - 2x - 3 = 0$

A. $x = \frac{3}{4}$ or $x = -\frac{1}{2}$. Factor the left side using unFOIL. The two coefficients of x in your final, factored equation need to have a product of 8, so first determine the factors of 8; you'll use $8 \cdot 1$ or $4 \cdot 2$. The product of the two constants in the binomials has to be 3, so you'll need $3 \cdot 1$. Through trial and error, you find that the 8 and 1 just don't work with this problem. Arranging the factors as $(4x \ 3)(2x \ 1)$, you see that the product of the two *Outer* terms is $4x$ and that the product of the two *Inner* terms is $6x$. The difference between $4x$ and $6x$ is $2x$, which is the middle term. Placing + and – in the correct positions, you have $8x^2 - 2x - 3 = (4x - 3)(2x + 1) = 0$. The multiplication property of zero (see Chapter 1) tells you that at least one of the binomials equals 0; $4x - 3 = 0$ gives you $x = \frac{3}{4}$, and $2x + 1 = 0$ gives you $x = -\frac{1}{2}$.

Q. Solve by factoring: $12x^2 + 44x + 24 = 0$

A. $x = -\frac{2}{3}$ or $x = -3$. First, factor a 4 out of each term to get $4(3x^2 + 11x + 6) = 0$. Now factor the trinomial to get $4(3x + 2)(x + 3) = 0$. Setting each binomial equal to 0, you get $x = -\frac{2}{3}$ and $x = -3$. Don't bother setting the 4 equal to 0, because $4 = 0$ is never true, so there's no solution from that factor.

5. Solve by factoring: $x^2 - 15x + 56 = 0$

6. Solve by factoring: $6x^2 + 19x + 15 = 0$

7. Solve by factoring: $x^2 + x - 30 = 0$

8. Solve by factoring: $5x^2 + 9x - 2 = 0$

9. Solve by factoring: $6x^2 - 45x - 24 = 0$

10. Solve by factoring: $4x^2 + 28x + 49 = 0$

11. Solve by factoring: $a^2x^2 - 2abx + b^2 = 0$

12. Solve by factoring: $25x^2 - 36 = 0$

Your Bag of Tricks: Factoring Multiple Ways

Factoring expressions in equations is a quick, efficient way of solving the equation, because you can obtain solutions by setting the individual factors equal to 0. However, using unFOIL isn't your only option for breaking down an equation.

Here's how to tackle quadratic and higher-degree equations that take more than one method of factoring:

✔ **Binomials:** When you have two terms in a quadratic equation or higher-power equation, look for

- A variable you can factor out, such as x^2 in $x^4 - x^2 = 0$
- The difference of squares — if $x^2 - a^2 = 0$, then $(x + a)(x - a) = 0$

✔ **Trinomials:** When you find three terms in the quadratic equation, look for

- A common factor
- A way to write the expression as the product of two binomials (unFOIL)

Q. Solve the equation incorporating two or more factoring methods: $2x^4 + 5x^3 - 3x^2 = 0$

A. $x = 0$, $x = \frac{1}{2}$, or $x = -3$. First factor x^2 out of each term to get $x^2(2x^2 + 5x - 3) = 0$. Then factor the trinomial in the parentheses to get $x^2(2x - 1)(x + 3) = 0$. Setting $x^2 = 0$, the solution is $x = 0$. When $2x - 1 = 0$, $x = \frac{1}{2}$. And when $x + 3 = 0$, $x = -3$.

Q. Solve the equation using two or more factoring methods: $5x^4 - 405 = 0$

A. $x = 3$ or $x = -3$. The two terms are divisible by 5, so, factoring, you get $5(x^4 - 81) = 0$. The binomial in the parentheses is the difference between two squares, which factors into $5(x^2 - 9)(x^2 + 9) = 0$. The first binomial factors, but the second doesn't, so you have $5(x - 3)(x + 3)(x^2 + 9) = 0$. Setting $x - 3 = 0$, you get $x = 3$; and when $x + 3 = 0$, $x = -3$; $x^2 + 9$ and 5 can't equal 0, so you have only two real answers.

13. Factor and solve for x: $5x^2 - 20 = 0$

14. Factor and solve for x: $x^5 - x^3 = 0$

15. Factor and solve for x: $2x^4 + 5x^3 - 3x^2 = 0$

16. Factor and solve for x: $400x^2 - 6{,}100x + 22{,}500 = 0$

Keeping Your Act Together: Factoring by Grouping

Grouping is a type of factoring in which you separate terms into collections that have the same factor. When grouping works, collecting terms and factoring puts common factors among the different collections or groups.

Q. Solve the equation by grouping the terms to factor: $x^3 + 3x^2 - 16x - 48 = 0$.

A. **$x = -3$, 4, or -4.** The first two terms have a common factor of x^2, and the second two terms have a common factor of -16. Factoring, you have $x^2(x + 3) - 16(x + 3) = 0$. The two new terms now have $(x + 3)$ as the common factor. Factor out that grouping to get $(x + 3)(x^2 - 16) = 0$. The second binomial is a quadratic that's the difference of two squares. Factoring that second binomial, $(x + 3)(x - 4)(x + 4) = 0$. Setting the factors equal to 0, you get $x = -3$, $x = 4$, and $x = -4$. (The equation $x^2 - 16 = 0$ can also be solved using the square root rule.)

17. Use factoring by grouping to solve:
$x^2 - 3x - 6x + 18 = 0$

18. Use factoring by grouping to solve:
$8x^3 + 12x^2 - 2x - 3 = 0$

19. Use factoring by grouping to solve:
$x^3 - 4x^2 - 25x + 100 = 0$

20. Use factoring by grouping to solve:
$x^3(x-2)^2 + 7x^2(x-2)^2 - x(x-2)^2 - 7(x-2)^2 = 0$

Resorting to the Quadratic Formula

The quadratic formula gives you the values of the variable x in the quadratic equation $ax^2 + bx + c = 0$. When using the formula, be sure to write the quadratic equation in exactly the correct format so that you correctly substitute the numbers for a, b, and c. People use the quadratic formula on quadratic equations that don't factor, but you can also use it on quadratic equations that do factor. Factoring is often quicker and more efficient, but sometimes the factorization just doesn't come to you. Having this fallback is nice.

The solutions of the quadratic equation $ax^2 + bx + c = 0$ are $x = \dfrac{-b \pm \sqrt{b^2 - 4ac}}{2a}$

When the quadratic formula yields a negative number under the radical, there's no real solution — no real number that x can be. Chapter 13 covers this type of situation. For now, when you have a negative under the radical, your answer is simply *no solution*.

Q. Use the quadratic formula to solve the equation $4x^2 + 7x - 5 = 0$.

A. $x = \dfrac{-7 + \sqrt{129}}{8} \approx 0.545$ or

$x = \dfrac{-7 - \sqrt{129}}{8} \approx -2.295$. The values of a, b, and c in the formula are 4, 7, and –5, respectively. Substituting into the formula,

$$x = \frac{-7 \pm \sqrt{7^2 - 4(4)(-5)}}{2(4)}$$

$$= \frac{-7 \pm \sqrt{49 + 80}}{8} = \frac{-7 \pm \sqrt{129}}{8}$$

So the values of x are $x = \dfrac{-7 + \sqrt{129}}{8} \approx 0.545$

and $x = \dfrac{-7 - \sqrt{129}}{8} \approx -2.295$.

Q. Use the quadratic formula to solve the equation $x^2 - 8x - 10 = 0$.

A. $x = 4 \pm \sqrt{26}$. The values of a, b, and c are 1, –8, and –10, respectively. Here's what you get when you substitute into the formula:

$$x = \frac{-(-8) \pm \sqrt{(-8)^2 - 4(1)(-10)}}{2(1)}$$

$$= \frac{8 \pm \sqrt{64 + 40}}{2} = \frac{8 \pm \sqrt{104}}{2} = \frac{8 \pm 2\sqrt{26}}{2}$$

$$= \frac{\overset{4}{\cancel{8}} \pm \cancel{2}\sqrt{26}}{\cancel{2}} = 4 \pm \sqrt{26}$$

The decimal equivalents for the solutions are $x \approx 9.099$ and $x \approx -1.099$.

21. Use the quadratic formula to solve $x^2 - 6x - 4 = 0$.

22. Use the quadratic formula to solve $2x^2 + 3x - 14 = 0$.

23. Use the quadratic formula to solve $3x^2 - 8x + 2 = 0$.

24. Use the quadratic formula to solve $x^2 + 4x - 14 = 0$.

Solving Quadratics by Completing the Square

The two most popular methods for solving a quadratic equation are factoring and using the quadratic formula. The last method you'd ever choose to use is completing the square. Why, you ask, do you have to know this method, then? That's a good question. And rather than say that it's good for you, I'll tell it like it is: The methods you use to solve quadratic equations by completing the square are the same methods you use to rewrite equations of conics in their standard form (which I cover in Chapter 10). Doing these rewrites here, in which you get the answer to a problem, is good practice. And convincing you that the method works for conics is easier if you've seen it before. Think of this technique as a small step toward a brighter future.

Here's how to solve the quadratic equation $ax^2 + bx + c = 0$ by completing the square:

1. **Divide every term in the equation by a, the coefficient of the x^2 term.**

2. **Move the constant term to the opposite side of the equation by adding or subtracting.**

3. **Find half the value of the coefficient on the x term; square the result of the halving; then add that amount to each side of the equation.**

4. **Factor the perfect square trinomial that you've created.**

    The factoring always has a binomial whose first term is an x and whose second term is half the coefficient of the x term on the left.

5. **Find the square root of each side of the equation.**

6. **Solve for the x by adding or subtracting a constant from each side.**

Q. Solve the quadratic equation $3x^2 + 4x - 2 = 0$ by completing the square.

A. $x = \dfrac{-2 \pm \sqrt{10}}{3}$. Using the steps outlined, first divide each term by 3:

$$\frac{3x^2}{3} + \frac{4x}{3} - \frac{2}{3} = 0$$

$$x^2 + \frac{4}{3}x - \frac{2}{3} = 0$$

Now add $\frac{2}{3}$ to each side to get

$$x^2 + \frac{4}{3}x = \frac{2}{3}.$$

Half of the coefficient on the x term, $\frac{4}{3}$, is $\frac{2}{3}$.

The square of $\frac{2}{3}$ is $\frac{4}{9}$. Add that square to both sides of the equation:

$$x^2 + \frac{4}{3}x + \frac{4}{9} = \frac{2}{3} + \frac{4}{9} = \frac{10}{9}.$$

Next, factor the trinomial on the left; you get the square of a binomial. (That's why the technique is called *completing the square*.) Then take the square root of each side:

$$x^2 + \frac{4}{3}x + \frac{4}{9} = \frac{10}{9}$$

$$\left(x + \frac{2}{3}\right)^2 = \frac{10}{9}$$

$$\sqrt{\left(x + \frac{2}{3}\right)^2} = \pm\sqrt{\frac{10}{9}}$$

$$x + \frac{2}{3} = \pm\frac{\sqrt{10}}{3}$$

Now subtract $\frac{2}{3}$ from each side to solve for x: $x = -\frac{2}{3} \pm \frac{\sqrt{10}}{3} = \frac{-2 \pm \sqrt{10}}{3}.$

---

25. Solve by completing the square: $x^2 - 8x - 9 = 0$

26. Solve by completing the square: $2x^2 + 7x - 4 = 0$

27. Solve by completing the square: $x^2 + 10x - 3 = 0$

28. Solve by completing the square: $ax^2 + bx + c = 0$

Working with Quadratic-Like Equations

When you solve equations with degrees higher than 2, you don't have the huge advantage of knowing a formula that always gives you the answer whether the terms factor or not. Because of this, you try to use every convenience to the fullest. One of these "conveniences" is recognizing a higher-degree equation that resembles a quadratic equation.

A *quadratic-like* equation has three terms, two with variables in them and one constant term. Also, the power on one of the variable terms is twice that of the other. The general format is: $ax^{2n} + bx^n + c = 0$. When you can, factor these equations as the product of two binomials using the same unFOIL format as with the quadratics (see the earlier section "UnFOILed Again! Successfully Factoring for Solutions").

Q. Solve the equation $8x^6 + 7x^3 - 1 = 0$.

A. $x = \frac{1}{2}$ or $x = -1$. Consider the quadratic equation that sort of looks like this equation: $8y^2 + 7y - 1 = 0$. This quadratic equation factors into $(8y - 1)(y + 1) = 0$. Use the same factorization pattern, replacing each y with x^3. You get $(8x^3 - 1)(x^3 + 1) = 0$. Using the multiplication property of zero, you determine that $8x^3 - 1 = 0$; $x = \sqrt[3]{\frac{1}{8}} = \frac{1}{2}$.

When you set the other factor equal to 0, you get $x^3 + 1 = 0$; $x = -1$.

You can do the same thing more formally with a process called *substitution*. You actually replace all the higher-degree terms with the lower and solve the new problem. The only trick is remembering to substitute back in when you're done.

Q. Solve the equation $x^4 + 4x^2 - 14 = 0$.

A. $x = \pm\sqrt{-2 + 3\sqrt{2}}$. First substitute y for x^2 and y^2 for x^4, giving you $y^2 + 4y - 14 = 0$. This quadratic equation requires the quadratic formula:

$$x^2 = \frac{-4 \pm \sqrt{4^2 - 4(1)(-14)}}{2(1)}$$

$$= \frac{-4 \pm \sqrt{16 + 56}}{2} = \frac{-4 \pm \sqrt{72}}{2}$$

$$= \frac{-4 \pm \sqrt{36}\sqrt{2}}{2} = \frac{-\overset{2}{4} \pm \overset{3}{6}\sqrt{2}}{\overset{}{2}}$$

$$= -2 \pm 3\sqrt{2}$$

You get two solutions, but wait! The solutions tell you what y is equal to, but $y = x^2$. Replacing the y with x^2 and solving for x using the square root property, you get

$$x^2 = -2 \pm 3\sqrt{2}$$

$$\sqrt{x^2} = \pm\sqrt{-2 \pm 3\sqrt{2}}$$

A negative under a radical isn't a real number, so you can use only the solutions $x = \pm\sqrt{-2 + 3\sqrt{2}}$.

29. Solve for x: $x^4 - 17x^2 + 16 = 0$

30. Solve for x: $x^8 - 15x^4 - 16 = 0$

31. Solve for x: $x^6 - 7x^3 - 8 = 0$

32. Solve for x: $x^8 - 626x^4 + 625 = 0$

33. Solve for x: $3x^4 - 22x^2 + 7 = 0$

34. Solve for x: $x^{10} + x^5 - 30 = 0$

Checking Out Quadratic Inequalities

You can solve a quadratic inequality such as $x^2 - 2x - 8 < 0$ most efficiently with a number line — sometimes called a *sign line*. (This method is also ideal for inequalities that involve higher-degree polynomials, and a number line also works when you have inequalities with fractions that have a polynomial in the numerator and denominator.)

First, write all the terms on one side of the inequality symbol and factor the expression. Then determine the *zeros* — the numbers that make the expression equal to 0. Put these zeros on a number line, from smallest to largest. The zeros determine the different intervals and divide into numbers where the expression is positive from where it's negative. You determine the sign in each interval by testing a number that falls between the zeros: Simply replace the x's in the factored expression with the test number. Then you can answer the question about whether the numbers in that interval are greater than or less than 0 (positive or negative).

Q. Solve the inequality $x^2 + 7x > 30$.

A. **$x < -10$ or $x > 3$.** First, subtract 30 from each side of the inequality: $x^2 + 7x - 30 > 0$. Factor the trinomial into $(x + 10)(x - 3) > 0$. The zeros of the factored form are $x = -10$ and $x = 3$. Put those zeros on a number line. Then determine whether the factored product is positive or negative in the three different intervals to the left of -10 and right of 3 and between those values. Test a number below -10, a number between -10 and 3, and a number above 3.

In Figure 2-1, you see that both factors are negative whenever x is a number smaller than -10. Multiply the two negative factors together, and that makes the product positive. All the numbers less than -10 will result in a positive product — which satisfies the inequality, because you want the result to be greater than 0. Back to the figure, the product of the two factors is negative between -10 and 3, so none of these numbers works. The numbers greater than 3 all work in the inequality, because both factors are positive. The solution for the inequality is $x < -10$ or $x > 3$. Writing this solution in interval notation, you get $(-\infty, -10)$, $(3, \infty)$. (See Chapter 1 for info on interval notation.)

Figure 2-1:
Signs indicate whether the factor is positive or negative.

$(x + 10)$ negative	$(x + 10)$ positive	$(x + 10)$ positive
$(x - 3)$ negative	$(x - 3)$ negative	$(x - 3)$ positive

$$\begin{array}{c|c|c} & -10 & 3 \end{array}$$

---

35. Solve the inequality: $x^2 + x - 42 \geq 0$

36. Solve the inequality: $x^3 - 9x < 0$

37. Solve the inequality: $x^2 + 4x + 7 \leq 0$

38. Solve the inequality: $\dfrac{x^2 - 4x - 5}{x + 2} \geq 0$

Answers to Problems on Quadratic (and Quadratic-Like) Equations and Inequalities

The following are the answers to the practice problems presented earlier in this chapter.

1 Solve for x: $4x^2 = 25$. The answer is $x = \pm\dfrac{5}{2}$.

Divide each side by 4. Then take the square root of each side. Both 25 and 4 are perfect squares, so the answer is a fraction — you don't have to find a decimal approximation, because the answer is a nice, rational number.

2 Solve for x: $3x^2 + 8 = 56$. The answer is $x = \pm 4$.

Subtract 8 from each side; then divide each side by 3. When you take the square root of each side, you get both +4 and –4.

3 Solve for x: $x^2 = 7$. The answer is $x = \pm\sqrt{7}$.

The number 7 isn't a perfect square, so leave the answer in *exact form* as plus or minus the square root of 7.

4 Solve for x: $5x^2 = 200$. The answer is $x = \pm 2\sqrt{10}$.

First, divide each side by 5 to get $x^2 = 40$. When you take the square root of each side, you have $\pm\sqrt{40}$, which you can simplify: $\pm\sqrt{40} = \pm\sqrt{4}\sqrt{10} = \pm 2\sqrt{10}$.

Use this rule to simplify radicals in which one of the factors of the number under the radical is a perfect square: $\sqrt{a \cdot b} = \sqrt{a}\sqrt{b}$.

5 Solve by factoring: $x^2 - 15x + 56 = 0$. The answer is $x = 7, 8$.

The trinomial factors into $(x - 7)(x - 8) = 0$. Setting the factors equal to 0, you get $x = 7$ or 8.

6 Solve by factoring: $6x^2 + 19x + 15 = 0$. The answer is $x = -\dfrac{5}{3}, -\dfrac{3}{2}$.

The trinomial factors into $(3x + 5)(2x + 3) = 0$. Set the factors equal to 0. Subtract and divide in each case to get the two solutions.

7 Solve by factoring: $x^2 + x - 30 = 0$. The answer is $x = -6, 5$.

The trinomial factors into $(x + 6)(x - 5) = 0$. Set the individual factors equal to 0 and solve for x in each case.

8 Solve by factoring: $5x^2 + 9x - 2 = 0$. The answer is $x = \dfrac{1}{5}, -2$.

The trinomial factors into $(5x - 1)(x + 2) = 0$. Set the factors equal to 0 and solve for x in each case.

9 Solve by factoring: $6x^2 - 45x - 24 = 0$. The answer is $x = -\dfrac{1}{2}, 8$.

The terms share a common factor of 3; factor it out first to get $3(2x^2 - 15x - 8) = 0$. Continue factoring to get $3(2x + 1)(x - 8) = 0$. Set each factor equal to 0 to get the answers. When you set the first factor equal to 0, you get $3 = 0$. This statement is never true, so you don't get a solution from that factor — just the other two.

10 Solve by factoring: $4x^2 + 28x + 49 = 0$. The answer is $x = -\dfrac{7}{2}$.

This equation has a perfect square trinomial on the left. Factored, you get $(2x + 7)^2 = 0$. Setting that factor equal to 0, you get the sole answer. This answer is called a double root, because the answer technically appears twice in the factorization: $(2x + 7)(2x + 7)$.

11 Solve by factoring: $a^2x^2 - 2abx + b^2 = 0$. The answer is $x = \dfrac{b}{a}$.

The trinomial factors as the square of the binomial $(ax - b)$. The trinomial is a perfect square trinomial. Set it equal to 0 to solve for the value of x you need.

12 Solve by factoring: $25x^2 - 36 = 0$. The answer is $x = \dfrac{6}{5}, -\dfrac{6}{5}$.

This equation includes the difference between two perfect squares, which factors into the difference and sum of the same two values. The binomial factors into $(5x - 6)(5x + 6) = 0$. Setting the two factors equal to 0, you get the two answers (two opposites).

13 Factor and solve for x: $5x^2 - 20 = 0$. The answer is $x = 2, -2$.

First, factor 5 out of each term to get $5(x^2 - 4) = 0$. The binomial in the parentheses factors into the difference and sum of the same two numbers, $5(x - 2)(x + 2) = 0$. Setting the factors equal to 0, you get the two solutions. Of course, setting $5 = 0$ doesn't yield a solution to the equation.

14 Factor and solve for x: $x^5 - x^3 = 0$. The answer is $x = 0, 1, -1$.

First, factor out the x^3 to get $x^3(x^2 - 1) = 0$. The binomial does factor, so you have $x^3(x - 1)(x + 1) = 0$. Setting the three factors equal to 0, you get the three solutions. Actually, the x^3 is a *triple root,* giving you the answer 0 three times.

People commonly miss the solution obtained by setting x or a power of x equal to 0 after factoring it out of the terms in an equation. Be careful to set $x^n = 0$ to get the answer that $x = 0$.

15 Factor and solve for x: $2x^4 + 5x^3 - 3x^2 = 0$. The answer is $x = 0, \dfrac{1}{2}, -3$.

First, factor out x^2 from each term to get $x^2(2x^2 + 5x - 3) = 0$. The quadratic in the parentheses factors, giving you $x^2(2x - 1)(x + 3) = 0$. Set the three factors equal to 0 to get the three solutions. Note that you have a double root at $x = 0$ because of the exponent on the x.

16 Factor and solve for x: $400x^2 - 6{,}100x + 22{,}500 = 0$. The answer is $x = \dfrac{25}{4}, 9$.

The numbers here seem enormous, but they're all divisible by 100. Factor that out first to get $100(4x^2 - 61x + 225) = 0$. The numbers used in the quadratic in the parentheses are also large, but think about the possible factors of each. The number 4 can be $4 \cdot 1$ or $2 \cdot 2$. The number 225 is 15^2, but it's also $225 \cdot 1$, $75 \cdot 3$, $45 \cdot 5$, and $25 \cdot 9$. It's this last pair, coupled with $4 \cdot 1$, that works to give a sum of 61 for the middle term. The quadratic factors to give you $100(4x - 25)(x - 9) = 0$. Set the two binomials equal to 0 and solve.

17 Use factoring by grouping to solve: $x^2 - 3x - 6x + 18 = 0$. The answer is $x = 3, 6$.

This equation is actually quadratic, so you can solve it by combining the two middle terms and then using unFOIL. To factor this expression by grouping, though, factor the x out of the first two terms and -6 out of the second two terms to get $x(x - 3) - 6(x - 3) = 0$. The grouping and factoring results in two terms — each with the common factor of $(x - 3)$. Now factor that binomial out of the two terms and you get $(x - 3)(x - 6) = 0$. Setting the factors equal to 0 gives you the two solutions.

18 Use factoring by grouping to solve: $8x^3 + 12x^2 - 2x - 3 = 0$. The answer is $x = -\dfrac{3}{2}, \dfrac{1}{2}, -\dfrac{1}{2}$.

The first two terms have a common factor of $4x^2$, and the second two terms have a common factor of -1 (you have to divide by -1 so that the binomials match). After doing the two factorizations, you end up with $4x^2(2x + 3) - 1(2x + 3) = 0$. Now, factoring out $(2x + 3)$, you write the equation as $(2x + 3)(4x^2 - 1) = 0$. The second binomial factors into the difference and sum of the same two numbers, $(2x + 3)(2x - 1)(2x + 1) = 0$. Set each factor equal to 0 to find the solutions.

19 Use factoring by grouping to solve: $x^3 - 4x^2 - 25x + 100 = 0$. The answer is $x = 4, 5, -5$.

Factor x^2 out of the first two terms and -25 out of the second two terms to get $x^2(x-4) - 25(x-4) = 0$. The common factor $(x-4)$ comes out of the two terms to give you $(x-4)(x^2-25) = 0$. The second binomial factors into the difference and sum of the same two numbers (or if you prefer, when you set the two binomials equal to 0, you can rewrite $x^2-25 = 0$ as $x^2 = 25$ and use the square root rule). Set each binomial equal to 0 and solve.

20 Use factoring by grouping to solve: $x^3(x-2)^2 + 7x^2(x-2)^2 - x(x-2)^2 - 7(x-2)^2 = 0$. The answer is $x = 2, -7, 1, -1$.

Notice that each of the four terms has a factor of $(x-2)^2$. After factoring that out, you can factor by grouping by taking x^2 out of the first two terms in the brackets and -1 out of the second two terms: $(x-2)^2[x^3 + 7x^2 - x - 7] = (x-2)^2[x^2(x+7) - 1(x+7)] = 0$. Factor out the $(x+7)$ within the brackets to get $(x-2)^2[(x+7)(x^2-1)] = (x-2)^2(x+7)(x-1)(x+1) = 0$. Set the individual factors equal to 0 to find the solutions. The solution $x = 2$ is a double root because of the power on its binomial.

21 Use the quadratic formula to solve $x^2 - 6x - 4 = 0$. The answer is $x = 3 \pm \sqrt{13}$.

The values of a, b, and c are 1, -6, and -4, respectively. Substituting these values into the quadratic formula, you get

$$x = \frac{-(-6) \pm \sqrt{(-6)^2 - 4(1)(-4)}}{2(1)}$$

$$= \frac{6 \pm \sqrt{36+16}}{2} = \frac{6 \pm \sqrt{52}}{2}$$

Be careful with the *order of operations* when simplifying the values under the radical. First, the $(-6)^2$ comes out to be a positive number. Note that I put the -6 in parentheses to emphasize that the negative sign is involved in the squaring. Second, be careful when multiplying the last three numbers. Count the number of negative signs — an even number of negative signs results in a positive product.

You can simplify the radical, because 52 is the product of a perfect square (4) and another number (13). After simplifying the radical, reduce the fraction:

$$x = \frac{6 \pm \sqrt{4}\sqrt{13}}{2} = \frac{6 \pm 2\sqrt{13}}{2}$$

$$= \frac{\overset{3}{\cancel{6}} \pm \cancel{2}\sqrt{13}}{\cancel{2}} = 3 \pm \sqrt{13}$$

22 Use the quadratic formula to solve $2x^2 + 3x - 14 = 0$. The answer is $x = 2, -\frac{7}{2}$.

The values of a, b, and c are 2, 3, and -14, respectively. Substituting these values into the quadratic formula, you get

$$x = \frac{-3 \pm \sqrt{3^2 - 4(2)(-14)}}{2(2)}$$

$$= \frac{-3 \pm \sqrt{9+112}}{4} = \frac{-3 \pm \sqrt{121}}{4}$$

$$= \frac{-3 \pm 11}{4}$$

Note that the answer doesn't contain a radical. The final answers are both rational numbers; you get them by first using the + and then the − in the fraction.

$$x = \frac{-3+11}{4} = \frac{8}{4} = 2 \quad \text{or} \quad x = \frac{-3-11}{4} = \frac{-14}{4} = -\frac{7}{2}$$

Having no radical in the answer means that you could have factored the quadratic to find the solution.

23 Use the quadratic formula to solve $3x^2 - 8x + 2 = 0$. The answer is $\boldsymbol{x = \frac{4 \pm \sqrt{10}}{3}}$.

The values of a, b, and c are 3, –8, and 2, respectively. Substituting these values into the quadratic formula, you get

$$x = \frac{-(-8) \pm \sqrt{(-8)^2 - 4(3)(2)}}{2(3)}$$

$$= \frac{8 \pm \sqrt{64 - 24}}{6} = \frac{8 \pm \sqrt{40}}{6}$$

Simplify the radical by rewriting it as the product of the root of a perfect square (4) times the root of the other factor (10). Then reduce the fraction:

$$x = \frac{8 \pm \sqrt{4}\sqrt{10}}{6} = \frac{8 \pm 2\sqrt{10}}{6}$$

$$= \frac{{}^{4}\cancel{8} \pm \cancel{2}\sqrt{10}}{{}^{3}\cancel{6}} = \frac{4 \pm \sqrt{10}}{3}$$

24 Use the quadratic formula to solve $x^2 + 4x - 14 = 0$. The answer is $\boldsymbol{x = -2 \pm 3\sqrt{2}}$.

Use 1, 4, and –14 for a, b, and c, respectively:

$$x^2 = \frac{-4 \pm \sqrt{4^2 - 4(1)(-14)}}{2(1)}$$

$$= \frac{-4 \pm \sqrt{16 + 56}}{2} = \frac{-4 \pm \sqrt{72}}{2}$$

$$= \frac{-4 \pm \sqrt{36}\sqrt{2}}{2} = \frac{-{}^{2}\cancel{4} \pm {}^{3}\cancel{6}\sqrt{2}}{\cancel{2}}$$

$$= -2 \pm 3\sqrt{2}$$

25 Solve by completing the square: $x^2 - 8x - 9 = 0$. The answer is $\boldsymbol{x = 9, -1}$.

The coefficient on the x^2 term is 1, so you can skip the first step. To solve the equation by completing the square, you first add 9 to each side and then add 16 to each side (16 is –4 squared — half the –8 coefficient squared). Write the terms on the left as the square of a binomial, and then take the square root of each side:

$$x^2 - 8x = 9$$

$$x^2 - 8x + 16 = 9 + 16$$

$$(x - 4)^2 = 25$$

$$\sqrt{(x - 4)^2} = \pm\sqrt{25}$$

$$x - 4 = \pm 5$$

Now solve for x by adding 4 to each side: $x = 4 \pm 5$. Obtain the two solutions by applying the + and the –: $x = 4 + 5 = 9$ and $x = 4 - 5 = -1$.

26 Solve by completing the square: $2x^2 + 7x - 4 = 0$. The answer is $x = \frac{1}{2}, -4$.

You first divide each term by 2 and then add 2 to each side to put the constant on the right side of the equation. Take half the coefficient on the x term and square it (half of $\frac{7}{2}$ is $\frac{1}{2} \cdot \frac{7}{2} = \frac{7}{4}$).

Then add that square to each side. Factor the terms on the left and simplify on the right, and then take the square root of each side:

$$\frac{2}{2}x^2 + \frac{7}{2}x - \frac{4}{2} = 0$$

$$x^2 + \frac{7}{2}x - 2 = 0$$

$$x^2 + \frac{7}{2}x = 2$$

$$x^2 + \frac{7}{2}x + \left(\frac{7}{4}\right)^2 = 2 + \left(\frac{7}{4}\right)^2$$

$$\left(x + \frac{7}{4}\right)^2 = \frac{32 + 49}{16} = \frac{81}{16}$$

$$\sqrt{\left(x + \frac{7}{4}\right)^2} = \pm\sqrt{\frac{81}{16}}$$

$$x + \frac{7}{4} = \pm\frac{9}{4}$$

Now subtract $\frac{7}{4}$ from each side and apply the + and – to get the answers:

$$x = -\frac{7}{4} \pm \frac{9}{4}$$

$$x = -\frac{7}{4} + \frac{9}{4} = \frac{2}{4} = \frac{1}{2}$$

or

$$x = -\frac{7}{4} - \frac{9}{4} = -\frac{16}{4} = -4$$

27 Solve by completing the square: $x^2 + 10x - 3 = 0$. The answer is $x = -5 \pm 2\sqrt{7}$.

Add 3 to each side of the equation. Then take half of 10 and square it — and add the 25 to each side. Factor on the left, and then take the square root of each side:

$$x^2 + 10x = 3$$

$$x^2 + 10x + 25 = 3 + 25$$

$$(x + 5)^2 = 28$$

$$\sqrt{(x + 5)^2} = \pm\sqrt{28}$$

$$x + 5 = \pm\sqrt{4}\sqrt{7} = \pm 2\sqrt{7}$$

Now solve for x by subtracting 5 from each side. The answer is $x = -5 \pm 2\sqrt{7}$.

28 Solve by completing the square: $ax^2 + bx + c = 0$. The answer is $x = \dfrac{-b \pm \sqrt{b^2 - 4ac}}{2a}$.

Look familiar? Well, it should. It's the quadratic formula. You can solve for the quadratic formula by taking the standard quadratic equation and solving for x by completing the square. You first divide each term by a. Subtract $\frac{c}{a}$ from each side. Then take half of the coefficient of the x term, $\frac{1}{2} \cdot \frac{b}{a} = \frac{b}{2a}$, square it, and add that squared value to both sides of the equation. Factor on the left and simplify on the right. Then take the square root of each side. Subtract $\frac{b}{2a}$ from each side and simplify:

$$ax^2 + bx + c = 0$$

$$\frac{a}{a}x^2 + \frac{b}{a}x + \frac{c}{a} = 0$$

$$x^2 + \frac{b}{a}x = -\frac{c}{a}$$

$$x^2 + \frac{b}{a}x + \left(\frac{b}{2a}\right)^2 = \left(\frac{b}{2a}\right)^2 - \frac{c}{a}$$

$$x^2 + \frac{b}{a}x + \frac{b^2}{4a^2} = \frac{b^2}{4a^2} - \frac{4ac}{4a^2}$$

$$\left(x + \frac{b}{2a}\right)^2 = \frac{b^2 - 4ac}{4a^2}$$

$$\sqrt{\left(x + \frac{b}{2a}\right)^2} = \pm\sqrt{\frac{b^2 - 4ac}{4a^2}}$$

$$x + \frac{b}{2a} = \pm\frac{\sqrt{b^2 - 4ac}}{\sqrt{4a^2}}$$

$$= -\frac{b}{2a} \pm \frac{\sqrt{b^2 - 4ac}}{2a}$$

$$= \frac{-b \pm \sqrt{b^2 - 4ac}}{2a}$$

29 Solve for x: $x^4 - 17x^2 + 16 = 0$. The answer is $x = 4, -4, 1, -1$.

Factor into the product of two quadratic binomials, $(x^2 - 16)(x^2 - 1) = 0$. Now, each binomial factors into the difference and sum of the same numbers, $(x - 4)(x + 4)(x - 1)(x + 1) = 0$. Set each factor equal to 0 for the solutions.

30 Solve for x: $x^8 - 15x^4 - 16 = 0$. The answer is $x = 2, -2$.

The left side factors into $(x^4 - 16)(x^4 + 1) = 0$. The first binomial factors, giving you $(x^2 - 4)$ $(x^2 + 4)$ $(x^4 + 1) = 0$. Only the first factor yields real solutions. When $x^2 - 4 = 0$, $x = 2$ or -2.

31 Solve for x: $x^6 - 7x^3 - 8 = 0$. The answer is $x = 2, -1$.

The trinomial factors into $(x^3 - 8)(x^3 + 1) = 0$. Setting the two binomials equal to 0, you get the two solutions after taking the cube root of each side.

32 Solve for x: $x^8 - 626x^4 + 625 = 0$. The answer is $x = 5, -5, 1, -1$.

First, factor the trinomial into $(x^4 - 625)(x^4 - 1) = 0$. Each binomial also factors: $(x^2 - 25)(x^2 + 25)$ $(x^2 - 1)(x^2 + 1) = 0$. Now factor the two binomials that are differences of perfect squares to get $(x - 5)(x + 5)(x^2 + 25)(x - 1)(x + 1)(x^2 + 1) = 0$. Four of the factors produce solutions when you set them equal to 0. The other factors yield answers that are complex numbers. (See Chapter 13 for the scoop on complex numbers.)

33 Solve for x: $3x^4 - 22x^2 + 7 = 0$. The answer is $x = \pm\frac{\sqrt{3}}{3}, \pm\sqrt{7}$.

First factor the trinomial into $(3x^2 - 1)(x^2 - 7) = 0$. The binomials don't factor, but you can find the solutions by setting the binomials equal to 0 and using the square root rule:

$$3x^2 - 1 = 0, \ 3x^2 = 1, \ x^2 = \frac{1}{3}, \ x = \pm\sqrt{\frac{1}{3}} = \pm\frac{\sqrt{3}}{3}$$

$$x^2 - 7 = 0, \ x^2 = 7, \ x = \pm\sqrt{7}$$

34 Solve for x: $x^{10} + x^5 - 30 = 0$. The answer is $\boldsymbol{x = -\sqrt[5]{6}, \sqrt[5]{5}}$.

Factor the left side into $(x^5 + 6)(x^5 - 5) = 0$. Setting the two factors equal to 0 and taking the fifth root of each side, you get the two solutions.

35 Solve the inequality: $x^2 + x - 42 \geq 0$. The answer is $\boldsymbol{x \leq -7 \text{ or } x \geq 6.}$

First, factor the quadratic into $(x + 7)(x - 6)$. When you set the factors equal to 0, you get the two zeros for the number line. Place the 7 and 6 on the number line and test numbers on either side and between the zeros to determine whether the values in the interval are positive or negative.

$(x+7)$ negative	$(x+7)$ positive	$(x+7)$ positive
$(x-6)$ negative	$(x-6)$ negative	$(x-6)$ positive

$$\overline{\qquad\qquad \underset{-7}{|} \qquad\qquad \underset{6}{|} \qquad\qquad}$$

The product of the two factors is positive when x is less than -7 or greater than 6. Include the two zeros in your answer. In interval notation, write the answer as $(-\infty, -7]$, $[6, \infty)$.

36 Solve the inequality: $x^3 - 9x < 0$. The answer is $\boldsymbol{x < -3 \text{ or } 0 < x < 3.}$

Factoring the expression on the left, you get $x(x - 3)(x + 3)$. Setting those factors equal to 0, the three zeros for the number line are 0, 3, and -3. None of these zeros appears in the answer, because you want values of x that make the expression negative. Place the zeros on the number line in order from the smallest to the largest. Then test the factors in the different intervals determined by the zeros.

x negative	x negative	x positive	x positive
$(x-3)$ negative	$(x-3)$ negative	$(x-3)$ negative	$(x-3)$ positive
$(x+3)$ negative	$(x+3)$ positive	$(x+3)$ positive	$(x+3)$ positive

$$\overline{\qquad \underset{-3}{|} \qquad\quad \underset{0}{|} \qquad\quad \underset{3}{|} \qquad}$$

The product is negative when an odd number of the factors is negative. The product of the three factors is negative when x is smaller than -3 and when x is between 0 and $+3$. In interval notation, write the answer as $(-\infty, -3), (0, 3)$.

37 Solve the inequality: $x^2 + 4x + 7 \leq 0$. **No solution.**

The quadratic doesn't factor, and when you use the quadratic formula, you get a negative under the radical. This tells you that there's no real solution. So, back to the drawing board. Just try some numbers for the x's in the expression on the left to see whether you can make it negative. When you let $x = -1$, you get $1 - 4 + 7 = 4$. That doesn't work. When you let $x = -10$, you get $100 - 40 + 7 = 67$. Might as well stop. No solutions make the quadratic 0, so the expression never changes signs.

38 Solve the inequality: $\dfrac{x^2 - 4x - 5}{x + 2} \geq 0$. The answer is **$-2 < x \leq -1$ or $x \geq 5$.**

Factoring the numerator of the fraction, you get $(x - 5)(x + 1)$. The two zeros from the numerator are 5 and -1, and the zero from the denominator is -2. Put those zeros on the number line and test to see what the signs of the factors are in the intervals.

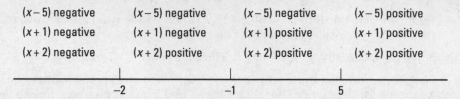

$(x - 5)$ negative	$(x - 5)$ negative	$(x - 5)$ negative	$(x - 5)$ positive
$(x + 1)$ negative	$(x + 1)$ negative	$(x + 1)$ positive	$(x + 1)$ positive
$(x + 2)$ negative	$(x + 2)$ positive	$(x + 2)$ positive	$(x + 2)$ positive

$$\underset{}{\overline{\qquad\qquad\underset{-2}{|}\qquad\qquad\underset{-1}{|}\qquad\qquad\underset{5}{|}\qquad\qquad}}$$

The result is positive when an even number of factors is negative. The fraction is positive between -2 and -1 and then, again, when x is greater than 5. You can include the zeros at 5 and -1, because they make the numerator equal to 0. That's allowed. You can't include the -2, because that makes the denominator equal to 0, and numbers divided by 0 are undefined. Written in interval notation, the solution is $(-2, -1]$, $[5, \infty)$.

Chapter 3

Rooting Out the Rational, the Radical, and the Negative

*R*ational numbers are those you can write as a fraction with an integer in both the numerator and denominator (but no 0 in the denominator). Rational numbers have the added attraction of having decimal equivalents that either terminate or repeat in a regular pattern. (This is what we call behaving rationally.)

Then you come to those pesky radicals. A *radical* indicates a root. The square root, cube root, fourth root, and so on are numbers whose repeated products give you the number under the radical sign. Numbers that are stuck under a radical sign — and have no exact value — don't behave very well. The decimal equivalents of numbers under a radical that aren't perfect squares or cubes, and so on, never end and never have a pattern that repeats. Consequently, numbers in this category are called *irrational numbers*.

You can use a negative exponent to show that you've moved a factor from the denominator of a fraction to the numerator. For instance, if you don't like the fraction $\frac{1}{2}$, you can move the 2 in the denominator up to the numerator by writing the $\frac{1}{2}$ as 2^{-1}. See! The fraction notation is gone! Writing and dealing with numbers and variables with negative exponents is often easier than working with fractions. The numbers with negative exponents and the same base combine more easily.

In this chapter, you see rational numbers, irrational numbers, and negative exponents. I put these exciting types of numbers into equations and then ask you to solve for any real solutions to the equations. You often need to use some special handling to solve these problems. Rest assured that the sections in this chapter can acquaint you with the rules involved.

Doing Away with Denominators with an LCD

What in the world is an LCD? It almost sounds like something out of the '60s. (Oh, no, that's LSD. Sorry!) Or maybe you're familiar with an LCD projector (liquid-crystal display). Wrong on both counts. In mathematics, an LCD is a *least common denominator*. When you have fractions and need to add or subtract them, you need a common denominator for the fractions. And of course, the most preferable is the *least* of these — you don't want the numbers to get too big.

A good process to use when solving equations that contain fractions is to change all the fractions so that they have the same denominator. Ideally, the common denominator is the LCD. So, what do you do after you go to all the trouble of changing all the fractions so they have the same denominator? You get rid of the denominators in the fractions by multiplying both sides of the equation by that LCD. The end result is a linear or quadratic or some other equation that you can solve more easily. The only catch is that you may introduce an extraneous solution.

An extraneous solution is an answer that doesn't work in the original equation. Extraneous solutions arise when you change the original equation to another, more-convenient form in order to solve the equation. Check for extraneous solutions by substituting any solutions you find back into the original equation.

Q. Solve for x in $\dfrac{3x}{x+2} - \dfrac{3}{x-1} = 1$ by finding an LCD and simplifying the equation.

A. $x = -\dfrac{1}{2}, 4$.

You first find the LCD of the fractions, which is their product, $(x + 2)(x - 1)$. Then multiply all the terms in the equation by this LCD:

$$\frac{3x}{x+2} \cdot \frac{(x+2)(x-1)}{1} - \frac{3}{x-1} \cdot \frac{(x+2)(x-1)}{1} = \frac{1}{1} \cdot \frac{(x+2)(x-1)}{1}$$

$$\frac{3x}{x+2} \cdot \frac{(x+2)(x-1)}{1} - \frac{3}{x-1} \cdot \frac{(x+2)(x-1)}{1} = \frac{1}{1} \cdot \frac{(x+2)(x-1)}{1}$$

$$3x(x-1) - 3(x+2) = (x+2)(x-1)$$

Multiply and distribute. Then set the equation equal to 0:

$$3x^2 - 3x - 3x - 6 = x^2 + x - 2$$

$$2x^2 - 7x - 4 = 0$$

This quadratic equation factors into $(2x + 1)(x - 4) = 0$. Setting the two factors equal to 0, you get $x = -\dfrac{1}{2}, 4$.

You need to check both answers. Neither creates a situation in which you have a 0 in the denominator, and both do work in the original equation. So, both answers work — neither solution is extraneous.

1. Solve for x: $\dfrac{x}{2} - \dfrac{3x}{4} = 2$

2. Solve for x: $\dfrac{4}{x+1} + \dfrac{6}{x-1} = 4$

3. Solve for x: $\dfrac{5}{x} - 6 = -\dfrac{3}{x+2}$

4. Solve for x: $\dfrac{2}{x-2} + \dfrac{8}{x+2} = \dfrac{8}{x^2-4}$

Simplifying and Solving Proportions

A proportion is an equation involving two ratios (fractions) set equal to each other.

The equation $\frac{14}{21} = \frac{24}{36}$ is a proportion. Both fractions in that proportion reduce to $\frac{2}{3}$, so it's fairly easy to see how this statement is true.

Proportions have some interesting, helpful, and easy-to-use properties:

In the proportion $\frac{a}{b} = \frac{c}{d}$

> ✔ **The cross-products are equal:** $a \cdot d = b \cdot c$.
>
> ✔ **The reciprocals are equal (you can flip the fractions):** $\frac{b}{a} = \frac{d}{c}$.
>
> ✔ **You can reduce the fractions vertically or horizontally:** You can divide out factors that are common to both numerators or both denominators or the left fraction or the right fraction. (You can't, however, divide out a factor from the numerator of one fraction and the denominator of the other.)

The properties of proportions come in useful when solving equations involving fractions. When you can, change an equation with fractions in it to a proportion for ease in solving.

Q. Solve the equation using the properties of proportions: $\frac{5}{x+3} + \frac{x}{10} = \frac{3x+6}{10}$.

A. $x = -8$ or $x = 2$. First, subtract $\frac{x}{10}$ from each side of the equation to create a proportion: $\frac{5}{x+3} + \frac{x}{10} - \frac{x}{10} = \frac{3x+6}{10} - \frac{x}{10}$

$$\frac{5}{x+3} = \frac{2x+6}{10}$$

Then reduce the right-hand fraction by factoring out a 2:

$$\frac{5}{x+3} = \frac{2(x+3)}{10}$$

$$\frac{5}{x+3} = \frac{\cancel{2}(x+3)}{\cancel{10}^{5}}$$

$$\frac{5}{x+3} = \frac{x+3}{5}$$

Now cross-multiply and simplify the resulting quadratic equation. Factor the quadratic and solve for the values of x:

$$5 \cdot 5 = (x+3)(x+3)$$
$$25 = x^2 + 6x + 9$$
$$0 = x^2 + 6x - 16$$
$$0 = (x+8)(x-2)$$
$$x = -8 \text{ or } x = 2$$

To check for extraneous solutions, you need to substitute the two different values in for x to see whether they work in the original equation. When $x = -8$, you get

$$\frac{5}{-8+3} + \frac{-8}{10} = \frac{3(-8)+6}{10}$$
$$\frac{5}{-5} + \frac{-4}{5} = \frac{-24+6}{10}$$
$$\frac{-5}{5} - \frac{4}{5} = \frac{-18}{10}$$
$$\frac{-9}{5} = \frac{-9}{5}$$

This solution works. Now, substituting in $x = 2$,

$$\frac{5}{2+3} + \frac{2}{10} = \frac{3(2)+6}{10}$$
$$\frac{5}{5} + \frac{1}{5} = \frac{6+6}{10}$$
$$\frac{6}{5} = \frac{12}{10}$$
$$\frac{6}{5} = \frac{6}{5}$$

Both solutions work.

5. Solve for x: $\dfrac{x+1}{9} = \dfrac{x+6}{24}$

6. Solve for x: $\dfrac{3x-2}{20} = \dfrac{12}{x+1} - \dfrac{7}{x+1}$

7. Solve for x: $\dfrac{144x}{x+71} = \dfrac{64(x+2)}{12(x+7)}$

8. Solve for x: $\dfrac{x+6}{5-2x} = \dfrac{x+2}{x-9}$

Wrangling with Radicals

Radical equations contain square roots or cube roots or some other roots of algebraic expressions. The most efficient way of solving these equations is to get rid of the radical signs. Squaring both sides once or twice gets rid of the square root radicals, but doing so may introduce an extraneous root. You need to check your solutions carefully.

Q. Solve for x: $\sqrt{5x-1}+2=x-1$

A. $x = 10$. First subtract 2 from each side to get the radical by itself on one side of the equation. Then square both sides of the equation:

$$\sqrt{5x-1}=x-3$$
$$\left(\sqrt{5x-1}\right)^2=(x-3)^2$$
$$5x-1=x^2-6x+9$$

Now set the quadratic equation equal to 0 and factor. Set the two factors equal to 0 to solve for x:

$$0=x^2-11x+10$$
$$=(x-10)(x-1)$$
$$x=10 \text{ or } x=1$$

To check the solutions in the *original* equation, first let $x = 10$, and you see that 10 is a solution:

$$\sqrt{5(10)-1}+2=10-1$$
$$\sqrt{50-1}+2=9$$
$$\sqrt{49}+2=9$$
$$7+2=9$$

Trying $x = 1$, you get

$$\sqrt{5(1)-1}+2=1-1$$
$$\sqrt{5-1}+2=0$$
$$\sqrt{4}+2=0$$
$$2+2\neq0$$

Thus, the solution $x = 1$ is extraneous; it's a solution of the quadratic equation, but it doesn't work in the original equation.

Q. Solve for x: $\sqrt{x+5}+\sqrt{2x+3}=9$

A. $x = 11$. This problem requires you to square the sides of the equation twice. You first move one of the radicals to the right and then square both sides. The result still has a radical in it, so you isolate that radical term on one side and square both sides again. Solve the resulting equation for x:

$$\sqrt{x+5}=9-\sqrt{2x+3}$$
$$\left(\sqrt{x+5}\right)^2=\left(9-\sqrt{2x+3}\right)^2$$
$$x+5=81-18\sqrt{2x+3}+2x+3$$
$$x+5=84+2x-18\sqrt{2x+3}$$
$$-x-79=-18\sqrt{2x+3}$$
$$(-x-79)^2=\left(-18\sqrt{2x+3}\right)^2$$
$$x^2+158x+6{,}241=324(2x+3)$$
$$x^2+158x+6{,}241=648x+972$$
$$x^2-490x+5{,}269=0$$
$$(x-11)(x-479)=0$$
$$x=11 \text{ or } x=479$$

Substituting in $x = 11$, you find that 11 is a solution:

$$\sqrt{11+5}+\sqrt{2(11)+3}=9$$
$$\sqrt{16}+\sqrt{25}=9$$
$$4+5=9$$

However, $x = 479$ is extraneous:

$$\sqrt{479+5}+\sqrt{2(479)+3}=9$$
$$\sqrt{484}+\sqrt{961}=9$$
$$22+31\neq9$$

9. Solve for x: $\sqrt{2x+5}-2=1$

10. Solve for x: $\sqrt{3x+7}+1=x$

11. Solve for x: $\sqrt{2x+7}-2x=1$

12. Solve for x: $\sqrt{2x-3}-\sqrt{x-2}=1$

Changing Negative Attitudes toward Negative Exponents

Negative exponents are really quite useful. They allow you to combine factors of numbers and variables with simple addition, subtraction, and multiplication. For instance, if you want to multiply $\frac{1}{x^3} \cdot \frac{3}{x^5} \cdot x^9$, just change the two denominators to numbers with negative exponents ($x^{-3} \cdot 3x^{-5} \cdot x^9$) and multiply the factors with the same base x by *adding the exponents* ($3x^{-3-5+9} = 3x^1$).

Here's the rule for changing fractions to numbers with negative exponents (and back):

$$\frac{1}{x} = x^{-1} \text{ and } \frac{1}{x^n} = x^{-n}$$

When negative exponents occur in equations, you can factor out the terms containing those negative exponents or treat those equations like other equations with a similar format. You usually want to get rid of the negative exponents (convert the terms with negative exponents to fractions) at the end of the problem to solve for the value of x.

Q. Solve for x: $x^{-1} + 3x^{-2} = 0$

A. $x = -3$. First factor out x^{-2} from each term, giving you $x^{-2}(x + 3) = 0$. Now set each factor equal to 0. The factor x^{-2} will never equal 0, because fractions (like $\frac{1}{x^2}$) can't equal 0 unless the numerator is 0. The solution for $x + 3 = 0$ is $x = -3$.

Q. Solve for x: $3x^{-2} + 2x^{-1} = 1$

A. $x = 3$ or $x = -1$. Rewrite the equation by setting it equal to 0: $3x^{-2} + 2x^{-1} - 1 = 0$. This gives you a quadratic-like equation (see Chapter 2 for more on this type of equation). The left side factors into the product of two binomials, $(3x^{-1} - 1)(x^{-1} + 1) = 0$. Setting the first factor equal to 0, you get

$$3x^{-1} - 1 = 0$$
$$\frac{3}{x} = 1$$
$$x = 3$$

The second factor yields you

$$x^{-1} + 1 = 0$$
$$\frac{1}{x} = -1$$
$$x = -1$$

Both of these solutions work in the original equation.

13. Solve for x: $x^{-2} = \frac{1}{9}$

14. Solve for x: $4x^{-2} - x^{-1} = 0$

15. Solve for x: $x^{-2} - 13x^{-1} + 36 = 0$

16. Solve for x: $8x^{-2} = 27x$

Divided Powers: Solving Equations with Fractional Exponents

You use fractional exponents to replace radicals and combinations of radicals and powers.

Here's how fractional exponents work:

$$\sqrt{a} = a^{1/2}, \text{ and more generally, } \sqrt[m]{a^n} = a^{n/m}$$

When solving equations with fractional exponents, use much the same technique as with negative exponents: Look for common factors, and look for quadratic-like or other equation patterns (see Chapter 2 for more info on equation patterns).

Q. Solve for x: $x^{5/3} + 2x^{4/3} - 3x = 0$

A. $x = 0$ or $x = -27$ or $x = 1$. Each term has a variable x in it, and the lowest power is x^1. Factor that out of each term, and then factor the trinomial using the same pattern as when factoring $y^2 + 2y - 3 = (y + 3)(y - 1) = 0$:

$$x^{5/3} + 2x^{4/3} - 3x = 0$$
$$x\left(x^{2/3} + 2x^{1/3} - 3\right) = 0$$
$$x\left(x^{1/3} + 3\right)(x^{1/3} - 1) = 0$$

Set each of the three factors equal to 0, and solve the three equations for x. When $x = 0$, you have a solution right there. When $x^{1/3} + 3 = 0$, subtract 3 from each side and then raise each side of the equation to the third power to get

$$x^{1/3} = -3$$
$$\left(x^{1/3}\right)^3 = (-3)^3$$
$$x = -27$$

Use the same technique with $x^{1/3} = 1$ to get that $x = 1$.

17. Solve for x: $x^{2/3} = 4$

18. Solve for x: $x^{2/3} - 13x^{1/3} + 36 = 0$

19. Solve for x: $x + x^{1/2} = 6$

20. Solve for x: $x^{1/2} = 4x^{1/6}$

Answers to Problems on Rooting Out the Rational, the Radical, and the Negative

The following are the answers to the practice problems presented in this chapter.

1 Solve for x: $\dfrac{x}{2} - \dfrac{3x}{4} = 2$. The answer is $x = -8$.

Clear the equation of fractions by multiplying each term by the LCD, which is 4:

$\dfrac{x}{2} \cdot \dfrac{4^2}{1} - \dfrac{3x}{4} \cdot \dfrac{4}{1} = 2 \cdot 4$. The equation simplifies to $2x - 3x = 8$. Combine the terms on the left, and the equation becomes $-x = 8$, or $x = -8$.

2 Solve for x: $\dfrac{4}{x+1} + \dfrac{6}{x-1} = 4$. The answer is $x = -\dfrac{1}{2}, 3$.

Multiply each fraction by the LCD, $(x + 1)(x - 1)$, to clear the equation of fractions. Then distribute the remaining values in each term:

$$\dfrac{4}{x+1} \cdot \dfrac{(x+1)(x-1)}{1} + \dfrac{6}{x-1} \cdot \dfrac{(x+1)(x-1)}{1} = 4 \cdot (x+1)(x-1)$$
$$4(x-1) + 6(x+1) = 4(x^2 - 1)$$
$$4x - 4 + 6x + 6 = 4x^2 - 4$$

Rewrite the equation as a quadratic set equal to 0, and factor for the solutions.

$$10x + 2 = 4x^2 - 4$$
$$0 = 4x^2 - 10x - 6$$
$$0 = 2(2x^2 - 5x - 3)$$
$$0 = 2(2x+1)(x-3)$$
$$x = -\dfrac{1}{2} \text{ or } x = 3$$

Both answers work — neither is extraneous (neither causes the denominator to be 0).

3 Solve for x: $\dfrac{5}{x} - 6 = -\dfrac{3}{x+2}$. The answer is $x = -\dfrac{5}{3}, 1$.

Multiply each fraction by the LCD, $x(x + 2)$. Then distribute the remaining values in each term:

$$\dfrac{5}{x} \cdot \dfrac{x(x+2)}{1} - 6 \cdot x(x+2) = -\dfrac{3}{x+2} \cdot \dfrac{x(x+2)}{1}$$
$$5(x+2) - 6x(x+2) = -3x$$
$$5x + 10 - 6x^2 - 12x = -3x$$

Rewrite the equation as a quadratic set equal to 0, and factor for the solutions. Neither answer is extraneous (neither causes the denominator to be 0):

$$10 - 6x^2 - 7x = -3x$$
$$0 = 6x^2 + 4x - 10$$
$$0 = 2(3x^2 + 2x - 5)$$
$$0 = 2(3x+5)(x-1)$$
$$x = -\dfrac{5}{3} \text{ or } x = 1$$

4 Solve for x: $\dfrac{2}{x-2} + \dfrac{8}{x+2} = \dfrac{8}{x^2-4}$. **No solution.**

Multiply each fraction by the LCD, $(x-2)(x+2) = x^2 - 4$. Then distribute the remaining values in each term. What's left is a linear equation that you can solve for x:

$$\frac{2}{\cancel{x-2}} \cdot \frac{\cancel{(x-2)}(x+2)}{1} + \frac{8}{x+2} \cdot \frac{(x-2)\cancel{(x+2)}}{1} = \frac{8}{\cancel{x^2-4}} \cdot \frac{\cancel{x^2-4}}{1}$$

$$2(x+2) + 8(x-2) = 8$$
$$2x + 4 + 8x - 16 = 8$$
$$10x - 12 = 8$$
$$10x = 20$$
$$x = 2$$

The solution $x = 2$ looks like a perfectly respectable answer. Unfortunately, it isn't. If you replace all the x's in the original equation with 2, you get 0 in the denominators of the first and last terms.

$$\frac{2}{2-2} + \frac{8}{2+2} = \frac{8}{2^2-4}$$
$$\frac{2}{0} + \frac{8}{4} = \frac{8}{0}$$

The value 2 is a solution of the linear equation that you created to solve the problem, but it's extraneous. It doesn't solve the original equation, and nothing else works in the original equation, either.

If you're skeptical about that *no solution* answer, just try another approach to the problem: Set the equation equal to 0 by moving the two terms to the right.

$$0 = \frac{8}{x^2-4} - \frac{2}{x-2} - \frac{8}{x+2}$$

Now combine them by finding a common denominator.

$$0 = \frac{8}{x^2-4} - \frac{2}{x-2} \cdot \frac{x+2}{x-2} - \frac{8}{x+2} \cdot \frac{x-2}{x+2} = \frac{8}{x^2-4} - \frac{2x+4}{x^2-4} - \frac{8x-16}{x^2-4}$$

$$= \frac{8-2(x+2)-8(x-2)}{x^2-4} = \frac{8-2x-4-8x+16}{x^2-4} = \frac{20-10x}{x^2-4} = \frac{10(2-x)}{x^2-4}$$

$$= \frac{-10\cancel{(x-2)}}{\cancel{(x-2)}(x+2)} = -\frac{10}{x+2}$$

The equation $0 = -\dfrac{10}{x+2}$ has no solution; the only way a fraction can equal 0 is if the numerator is 0.

5 Solve for x: $\dfrac{x+1}{9} = \dfrac{x+6}{24}$. The answer is $x = 2$.

You can first simplify the proportion by reducing horizontally across the bottom; both denominators are divisible by 3. Then cross-multiply, distribute the factors, and solve the resulting linear equation:

$$\frac{x+1}{\cancel{9}_3} = \frac{x+6}{\cancel{24}_8}$$
$$8(x+1) = 3(x+6)$$
$$8x + 8 = 3x + 18$$
$$5x = 10$$
$$x = 2$$

6 Solve for x: $\dfrac{3x-2}{20} = \dfrac{12}{x+1} - \dfrac{7}{x+1}$. The answer is $x = \dfrac{17}{3}, -6$.

Combine the two terms on the right with the same denominator by performing the subtraction. Then cross-multiply and rewrite the resulting equation as a quadratic set equal to 0. Factor the quadratic and solve for x:

$$\frac{3x-2}{20} = \frac{12}{x+1} - \frac{7}{x+1}$$
$$\frac{3x-2}{20} = \frac{5}{x+1}$$
$$(3x-2)(x+1) = 100$$
$$3x^2 + x - 2 = 100$$
$$3x^2 + x - 102 = 0$$
$$(3x - 17)(x + 6) = 0$$
$$x = \frac{17}{3} \text{ or } x = -6$$

Both solutions work — neither is extraneous.

7 Solve for x: $\dfrac{144x}{x+71} = \dfrac{64(x+2)}{12(x+7)}$. The answer is $x = -\dfrac{71}{13}, 1$.

You can reduce the fraction on the right by dividing the numerator and denominator by 4. Then you can reduce horizontally, across the top, by dividing both numerators by 16. Cross-multiply what's left and simplify to form a quadratic equation you can set equal to 0 and factor for the solution:

$$\frac{144x}{x+71} = \frac{\overset{16}{\cancel{64}}(x+2)}{\underset{3}{\cancel{12}}(x+7)}$$
$$\frac{\overset{9}{\cancel{144}}x}{x+71} = \frac{\overset{1}{\cancel{16}}(x+2)}{3(x+7)}$$
$$9x(3)(x+7) = (x+71)(x+2)$$
$$27x^2 + 189x = x^2 + 73x + 142$$
$$26x^2 + 116x - 142 = 0$$
$$2(13x^2 + 58x - 71) = 0$$
$$2(13x + 71)(x - 1) = 0$$
$$x = -\frac{71}{13} \text{ or } x = 1$$

8 Solve for x: $\dfrac{x+6}{5-2x} = \dfrac{x+2}{x-9}$. The answer is $x = \dfrac{16}{3}, -4$.

Cross-multiply and then simplify the terms by distributing. Rewrite the equation as a quadratic equation set equal to 0. Factor and find for the solutions:

$$\frac{x+6}{5-2x} = \frac{x+2}{x-9}$$
$$(x+6)(x-9) = (x+2)(5-2x)$$
$$x^2 - 3x - 54 = -2x^2 + x + 10$$
$$3x^2 - 4x - 64 = 0$$
$$(3x - 16)(x + 4) = 0$$
$$x = \frac{16}{3} \text{ or } x = -4$$

9 Solve for x: $\sqrt{2x+5}-2=1$. The answer is $x = 2$.

Add 2 to each side of the equation, and then square both sides, which gives a linear equation. Subtract 5 from each side and solve for x. The answer is $x = 2$.

$$\sqrt{2x+5}-2=1$$
$$\sqrt{2x+5}=3$$
$$\left(\sqrt{2x+5}\right)^2=3^2$$
$$2x+5=9$$
$$2x=4$$
$$x=2$$

Substituting this number into the original equation, you end up with $3 - 2 = 1$, so the answer checks.

10 Solve for x: $\sqrt{3x+7}+1=x$. The answer is $x = 6$.

Subtract 1 from each side, and then square both sides of the equation to get a quadratic. Set the equation equal to 0 and factor.

$$\sqrt{3x+7}+1=x$$
$$\sqrt{3x+7}=x-1$$
$$\left(\sqrt{3x+7}\right)^2=(x-1)^2$$
$$3x+7=x^2-2x+1$$
$$0=x^2-5x-6$$

$0 = (x - 6)(x + 1) = 0$. When you try $x = 6$ in the original equation, you get $5 + 1 = 6$. When you try $x = -1$ in the original equation, you get $2 + 1 = -1$, which isn't true; so $x = -1$ is an extraneous solution.

11 Solve for x: $\sqrt{2x+7}-2x=1$. The answer is $x = 1$.

Add $2x$ to each side of the equation, and then square both sides to get a quadratic.

$$\sqrt{2x+7}-2x=1$$
$$\sqrt{2x+7}=2x+1$$
$$\left(\sqrt{2x+7}\right)^2=(2x+1)^2$$
$$2x+7=4x^2+4x+1$$

Set this quadratic equation equal to 0 and factor:

$$0=4x^2+2x-6$$
$$0=2(2x^2+x-3)$$
$$0=2(2x+3)(x-1)$$

Setting the factors equal to 0, the first solution, $-\frac{3}{2}$, doesn't work, because it makes the original equation read $2 + 3 = 1$. The second solution, $x = 1$, does work; you get $3 - 2 = 1$.

12 Solve for x: $\sqrt{2x-3} - \sqrt{x-2} = 1$. The answer is $x = $ **6, 2.**

You have to square both sides of this equation twice to get rid of the radicals. First, add $\sqrt{x-2}$ to each side, and then square both sides. Simplify the terms, and then isolate the remaining radical expression by putting the rest of the terms on the other side. Square both sides, and then solve the resulting quadratic equation:

$$\sqrt{2x-3} = 1 + \sqrt{x-2}$$
$$\left(\sqrt{2x-3}\right)^2 = \left(1 + \sqrt{x-2}\right)^2$$
$$2x-3 = 1 + 2\sqrt{x-2} + x - 2$$
$$2x-3 = -1 + x + 2\sqrt{x-2}$$
$$x-2 = 2\sqrt{x-2}$$
$$(x-2)^2 = \left(2\sqrt{x-2}\right)^2$$
$$x^2 - 4x + 4 = 4(x-2)$$
$$x^2 - 4x + 4 = 4x - 8$$
$$x^2 - 8x + 12 = 0$$
$$(x-6)(x-2) = 0$$
$$x = 6 \text{ or } x = 2$$

Both the 6 and the 2 work in the original equation.

13 Solve for x: $x^{-2} = \frac{1}{9}$. The answer is $x = $ **3, –3.**

First rewrite the equation replacing the negative exponent: $\frac{1}{x^2} = \frac{1}{9}$. You have a proportion, so you can "flip" the fractions and set the reciprocals equal to one another. (If you need a refresher on proportion properties, refer to the section "Simplifying and Solving Proportions," earlier in this chapter.) The new equation reads: $x^2 = 9$. Use the square root property (take the square root of both sides), and you get $x = \pm 3$.

14 Solve for x: $4x^{-2} - x^{-1} = 0$. The answer is $x = $ **4.**

Factor x^{-2} out of each term to get $x^{-2}(4 - x) = 0$. Setting x^{-2} equal to 0 doesn't yield you an answer, because that equation has no solution. The only way a fraction can equal 0 is if the numerator equals 0, and with a 1 in the numerator of this fraction, that can never happen. The other factor gives you the solution $x = 4$.

15 Solve for x: $x^{-2} - 13x^{-1} + 36 = 0$. The answer is $x = \frac{1}{9}, \frac{1}{4}$.

This trinomial can be considered to be *quadratic-like* (see Chapter 2), because the power on the first term is twice that of the second and the last term is a constant. Factoring the trinomial into the product of two binomials, you get $(x^{-1} - 9)(x^{-1} - 4) = 0$. Setting these factors equal to 0, use the equations $\frac{1}{x} - 9 = 0$ or $\frac{1}{x} - 4 = 0$ to solve for the solutions:

$$\frac{1}{x} - 9 = 0 \text{ or } \frac{1}{x} - 4 = 0$$

$$\frac{1}{x} = 9 \text{ or } \frac{1}{x} = 4$$

$$\frac{1}{x} = \frac{9}{1} \text{ or } \frac{1}{x} = \frac{4}{1}$$

$$\frac{x}{1} = \frac{1}{9} \text{ or } \frac{x}{1} = \frac{1}{4}$$

16 Solve for x: $8x^{-2} = 27x$. The answer is $\boldsymbol{x = \dfrac{2}{3}}$.

Subtract $27x$ from each side and then factor x^{-2} out of each term to get $x^{-2}(8 - 27x^3) = 0$. The first factor, x^{-2}, doesn't yield a solution, but setting the second factor equal to 0, you get

$$8 = 27x^3$$
$$\frac{8}{27} = x^3$$
$$\sqrt[3]{\frac{8}{27}} = x$$
$$\frac{2}{3} = x$$

TIP

When you have only two terms in an equation with negative exponents, consider rewriting the equation as a proportion. You can do problems 14 and 16 this way. For problem 14, you write the equation as

$$4x^{-2} = x^{-1}$$
$$\frac{4}{x^2} = \frac{1}{x}$$

Then cross-multiply. The downside to this method is that you're apt to introduce extraneous solutions, so you have to be especially wary. Likewise, problem 16 can appear as $\dfrac{8}{x^2} = \dfrac{27x}{1}$. Simply cross-multiply and solve for x.

17 Solve for x: $x^{2/3} = 4$. The answer is $\boldsymbol{x = 8, -8.}$

First, cube each side of the equation (raise each side to the third power): $(x^{2/3})^3 = (4)^3$. That gives you the equation $x^2 = 64$. Now use the square root property, and you end up with $x = 8$ or $x = -8$.

18 Solve for x: $x^{2/3} - 13x^{1/3} + 36 = 0$. The answer is $\boldsymbol{x = 64, 729.}$

This trinomial factors using the same factorization pattern as in $y^2 - 13y + 36 = (y - 4)(y - 9) = 0$. The factorization with the fractional exponents is $(x^{1/3} - 4)(x^{1/3} - 9) = 0$. Setting each factor equal to 0 and solving by cubing both sides of the equations, you get $(x^{1/3})^3 = 4^3$, $x = 64$, and $(x^{1/3})^3 = 9^3$, $x = 729$.

19 Solve for x: $x + x^{1/2} = 6$. The answer is $\boldsymbol{x = 4.}$

Rearrange the equation to equal 0, giving you the trinomial $x + x^{1/2} - 6 = 0$. Next, factor the trinomial into $(x^{1/2} - 2)(x^{1/2} + 3) = 0$. Setting the first factor equal to 0 and solving for x, you get $x^{1/2} - 2 = 0$, $x^{1/2} = 2$, $(x^{1/2})^2 = 2^2$, giving you that $x = 4$. But when you try to solve $x^{1/2} + 3 = 0$, you get $x^{1/2} = -3$, which is an impossible statement; the only answer is 4.

20 Solve for x: $x^{1/2} = 4x^{1/6}$. The answer is $\boldsymbol{x = 0, 64.}$

First, rearrange the equation to equal 0, and then factor out the common factor of $x^{1/6}$.

$$x^{1/2} - 4x^{1/6} = 0$$
$$x^{1/6}\left(x^{1/3} - 4\right) = 0$$

When you set $x^{1/6} = 0$, you get $x = 0$ after raising each side to the sixth power. When setting $x^{1/3} - 4 = 0$, move the 4 to the right and cube each side to get $x = 64$.

Chapter 4

Graphing for the Good Life

● ●

In This Chapter

▶ Getting situated with points, axes, coordinates, quadrants, symmetry, and intercepts

▶ Graphing lines every which way but loose

▶ Recognizing what you need when graphing polynomials, radicals, and absolute value equations

▶ Becoming familiar with graphing calculator capabilities

● ●

A graph is a mathematical picture. In statistics or other topics that deal with numbers of things, you see bar graphs, scatter plots, line graphs, and so on. In algebra, you use graphs to represent all the sets of numbers that work in a particular equation. A graph is very useful — it gives you visual information right upfront. For instance, it can tell you how high or how low the points go. It can show you, quickly, if the graph represents points that spread all over the place or points that are limited in their scope.

In this chapter, you become familiar with the coordinate graphing system and its many facets. Your goal is to quickly graph equations, so in this chapter, you see how to use the characteristics of the different types of equations in order to make the process easier and faster.

I also provide a bit of information on graphing with some technological help. The graphing calculator information near the end of the chapter is purposely generic so you can see the technological graphing possibilities and then go to your particular calculator manual for all the gory details.

Coordinating Axes, Coordinates of Points, and Quadrants

The *Cartesian coordinate system* is the standard graphing arrangement set up for mathematical points and equations. In this system, the *x*- and *y*-axes (singular: axis) are perpendicular lines that divide a plane into four sections called *quadrants*. The *origin* is the point at the center of the plane where the axes intersect. *Coordinates* are ordered pairs of numbers, such as (–3, 4) or (7, 0); any ordered pair (*x*, *y*) tells you how many units from the origin a particular point is in the *x* direction (left or right) and the *y* direction (up or down). Thus, a point on one of the axes always has a 0 for one of the coordinates, and the origin has a 0 for both coordinates. Figure 4-1 shows you the coordinate system's axes, origin, quadrants, and a few select points.

The quadrants are labeled in a counterclockwise direction, starting with Quadrant I in the upper-right corner. Given which quadrant a point is in, you can tell whether the point's *x*- and *y*-values are positive or negative. A negative coordinate indicates that you move the point left or down from the origin; positive coordinates go to the right or upward. Therefore, all the points in a particular quadrant have the same signs on their *x*- and *y*-coordinates. For instance, all points in Quadrant IV have a positive *x*-coordinate and a negative *y*-coordinate.

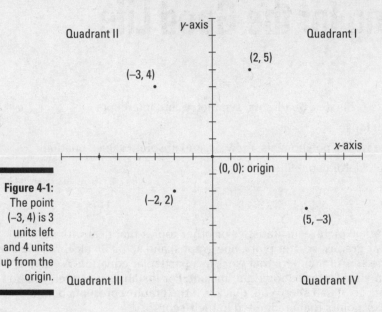

Figure 4-1:
The point (–3, 4) is 3 units left and 4 units up from the origin.

Q. Which quadrant do the points (–3, 4), (–1, 6) and (–5, 5) lie in?

A. **Quadrant II.** The *x*-coordinates are all negative, so that means that the points are in either Quadrant II or Quadrant III. The *y*-coordinates are positive, which is true of points in Quadrant I or Quadrant II. The only quadrant satisfying both of these conditions is Quadrant II.

Use Figure 4-2 to answer questions 1 through 4.

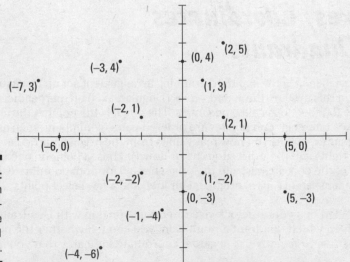

Figure 4-2:
Points in their quad-rants for use in problems 1 through 4.

1. Name the points in Figure 4-2 that lie in Quadrant II or on the *x*-axis.

2. Name the points in Figure 4-2 that lie in Quadrant IV or on the *y*-axis.

3. Identify the quadrant that contains all points with a negative *x*-coordinate and a negative *y* coordinate.

4. If you draw a segment connecting points (–4, 4) and (2, –2), what point lies on that line segment that has *x*- and *y*-coordinates with the same value (the numbers are the same)?

Crossing the Line: Using Intercepts and Symmetry to Graph

The *intercepts* of a graph are where the graph of an equation crosses an axis. The *x*-intercepts all have the general form (*h*, 0) — the *x*-coordinate is some number, and the *y*-coordinate is 0. The *y*-intercepts all have the general form (0, *k*), where the *x*-coordinate is 0 and the *y*-coordinate is some number. Replacing one of the variables in an equation with a 0 makes the equation simpler, so finding intercepts by putting in 0s also simplifies the graphing.

To find the *x*-intercepts of the graph of an equation, let *y* = 0 and solve for *x*. (If you have the *y* isolated on one side of the equation, you solve for *x*-intercepts the same way you'd solve any equation set equal to 0 — the *x*-intercepts are *solutions* or *roots* of the equation.) To find the *y*-intercepts, let *x* = 0 and solve for *y*.

A graph has *symmetry* if half of it looks like a mirror reflection or 180-degree rotation of the other half. Think of the letter *V*, which has symmetry about a vertical line through its center, or the letter *S*, which has rotational symmetry about its middle. Instead of plotting points to check whether a graph is symmetrical, you can try the following algebraic tests.

Here are the types of symmetry and how to look for them:

- A graph is *symmetric with respect to the y-axis* if the graph is a mirror image on either side of that vertical (y) axis. If you replace all the x's in the equation with $-x$, the equation remains unchanged.

- A graph is *symmetric with respect to the x-axis* if it's a mirror image on either side of the horizontal (x) axis. If you replace all the y's in the equation with $-y$, the equation remains unchanged.

- A graph is *symmetric about the origin* if a 180-degree turn around the origin makes the parts of the graph coincide. If you replace all the x's with $-x$ and the y's with $-y$, the equation remains unchanged.

Q. Use intercepts and symmetry to help graph the equation $x^2 + (y - 3)^2 = 25$.

A. Letting $y = 0$, you get $x^2 + 9 = 25$, so $x^2 = 16$ and $x = \pm 4$. The two x-intercepts are $(4, 0)$ and $(-4, 0)$. When you let $x = 0$, you get $(y - 3)^2 = 25$. Taking the square root of each side, $y - 3 = \pm 5$, so $y = 3 + 5 = 8$ or $y = 3 - 5 = -2$. The two y-intercepts are $(0, 8)$ and $(0, -2)$. When you replace x with $-x$ in the equation, the original equation doesn't change, so the graph is symmetric

with respect to the y-axis. The graph in Figure 4-3a shows the four intercepts graphed on the coordinate axes. You may recognize this equation as the standard form for a circle (see Chapter 10). Even if you don't immediately recognize that, you can just plug and plot a few more points and use the fact that the graph is symmetric about the y-axis to fill in the rest of the graph. Figure 4-3b shows the whole graph drawn in.

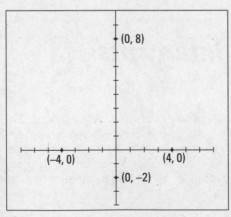
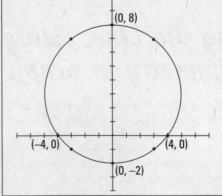

Figure 4-3: The four intercepts drawn in and the rest of the graph, showing that the equation is a circle.

a

b

5. Use intercepts and symmetry to sketch the graph of $y = x^2 - 9$.

6. Use intercepts and symmetry to sketch the graph of $y = 3 - |x|$.

7. Use intercepts and symmetry to sketch the graph of $y = \dfrac{1}{x^2}$.

8. Use intercepts and symmetry to sketch the graph of $y = \sqrt{16 - x^2}$.

Graphing Lines Using Slope-Intercept and Standard Forms

Just two points determine a line; only one distinct line can go through those two points. People usually describe a line with one of three common forms of the equation: *slope-intercept, standard,* or *point-slope.*

- **Slope-intercept form:** $y = mx + b$. The m is a constant that equals the slope of the line, and the b is the y-coordinate of the y-intercept of the line.

- **Standard form:** $Ax + By = C$. When using this form, the slope is equal to $-\frac{A}{B}$, the x-intercept is $\left(\frac{C}{A}, 0\right)$, and the y-intercept is $\left(0, \frac{C}{B}\right)$.

- **Point-slope form:** $y - y_1 = m(x - x_1)$. The m is the constant that equals the slope of the line, and (x_1, y_1) are the coordinates of a point on the line.

Q. Use the equation of the line $4x - 3y = 12$ to find the slope and two intercepts of the line. Then change this equation to slope-intercept form and verify the slope and y-intercept. Sketch a graph of this line.

A. The slope is $\frac{4}{3}$, the x-intercept is $(3,0)$, and the y-intercept is $(0, -4)$.

You get these answers from the fractions I give you in the second bullet point.

slope: $\frac{A}{B} = -\frac{4}{-3} = \frac{4}{3}$

x-intercept: $\left(\frac{C}{A}, 0\right) = \left(\frac{12}{4}, 0\right) = (3, 0)$

y-intercept: $\left(0, \frac{C}{B}\right) = \left(0, \frac{12}{-3}\right) = (0, -4)$

To write this standard-form equation in slope-intercept form, solve for y: subtract $4x$ from each side and then divide each term by -3.

$$4x - 3y = 12$$
$$-3y = -4x + 12$$
$$\frac{-3y}{-3} = \frac{-4x}{-3} + \frac{12}{-3}$$
$$y = \frac{4}{3}x - 4$$

You see that the slope, $\frac{4}{3}$, and y-intercept, -4, match what you found earlier. To sketch the line, you can use either the two intercepts or the y-intercept and the slope. To use the y-intercept and slope, start at the y-intercept and look at the denominator of the fraction representing the slope (if the slope is an integer, then the denominator is 1). The denominator tells you how many units to move to the right from the y-intercept. Then look at the numerator of the slope. Count up as many units as are in the numerator if the slope is positive; count down if it's negative. Check out Figure 4-4 to see how to graph this line using both methods.

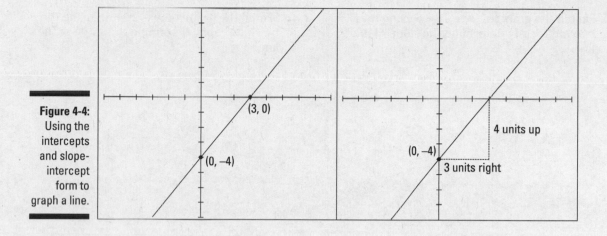

Figure 4-4:
Using the intercepts and slope-intercept form to graph a line.

9. Sketch the graph of $y = \frac{2}{3}x - 4$ using the slope and y-intercept.

10. Sketch the graph of $y = -5x + 1$ using the slope and y-intercept.

11. Sketch the graph of $2x + 3y = 6$ using the two intercepts; determine the slope of the line.

12. Sketch the graph of $9x - 2y = 18$ using the two intercepts; determine the slope of the line.

Graphing Basic Polynomial Curves

A *polynomial* is a smooth curve that has x-values ranging from negative infinity to positive infinity. The equation representing the curve determines the y-values for points on the curve. There's just one y-value for every x-value, because polynomials are types of functions.

A *function* is a special type of relationship or equation in which you find no more than one y-value for every x-value.

The highest degree (exponent) of a polynomial tells you the maximum number of x-intercepts the polynomial can have and the maximum number of *turning points* (where the graph changes direction from going upward to downward or vice versa). If the highest degree of the polynomial is n, then you will find a maximum of n x-intercepts and a maximum of $n - 1$ turning points. You use the intercepts, turning points, symmetry, and a few extra plug-and-plot points to sketch in the graph of the polynomial.

A polynomial may have fewer intercepts or turning points but never more than the maximum (makes sense, huh?). A polynomial has fewer x-intercepts if the equation has a double or triple root — the same solution appearing more than once. You also find fewer x-intercepts when roots are imaginary. Refer to Chapter 13 for information on imaginary and complex numbers. You can find out more about graphing and working with polynomials in Chapter 7.

Q. Sketch the graph of the polynomial $y = x^3 - 9x$.

A. Let $y = 0$ to solve for the x-intercepts: $0 = x^3 - 9x = x(x^2 - 9) = x(x - 3)(x + 3)$. The x-intercepts are $(0, 0)$, $(3, 0)$, and $(-3, 0)$. When you let $x = 0$ to find the y-intercept, you get $(0, 0)$, also an x-intercept. This repeated intercept always happens when one of the intercepts is the origin. This graph is symmetric with respect to the origin, because if you replace each x with $-x$ and each y with $-y$, you get $(-y) = (-x)^3 - 9(-x)$. This equation simplifies to $-y = -x^3 + 9x$. Multiplying each side of the equation by -1, you get the original equation — there was no change. (See the previous section, "Crossing the Line: Using Intercepts and Symmetry to Graph," for more info on symmetry tests.) Figure 4-5 shows the intercepts and a few other points plotted. These points should be enough, along with the symmetry, to sketch in the whole curve. You can see that this graph contains two turning points.

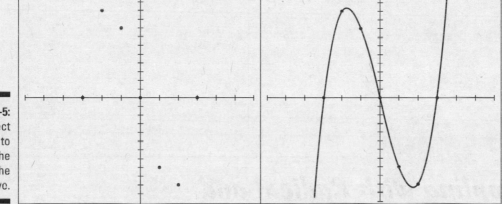

Figure 4-5: Connect the dots to sketch the graph of the curve.

13. Use intercepts, turning points, symmetry, and a few select points to sketch the graph of $y = x^3 - 4x^2$.

14. Use intercepts, turning points, symmetry, and a few select points to sketch the graph of $y = x^4 - x^2$.

15. Use intercepts, turning points, symmetry, and a few select points to sketch the graph of $y = x(x + 3)(x - 4)(x - 6)$.

16. Use intercepts, turning points, symmetry, and a few select points to sketch the graph of $y = (x^2 - 36)(x^2 - 25)$.

Grappling with Radical and Absolute Value Functions

Radical equations and absolute value functions have distinctive-looking curves. Radical equations are usually restricted in their domains (possible x-values — see Chapter 5), because negatives under even-powered radicals represent numbers that don't exist. Consequently, you usually see nice, smoothly rising (or falling) curves that start or end abruptly at one point in their domain. Absolute value functions have a distinctive V shape to them, facing either upward or downward.

Intercepts and symmetry help with the sketches of these curves, too, but even more helpful are the "endpoint" in the case of the radical function and the tip of the V in the case of an absolute value function.

Q. Sketch the graph of $y = \sqrt{x^2 - 9}$.

A. The domain of this function doesn't include any numbers between –3 and 3. Any x between those two numbers gives you a negative value under the radical. The two x-intercepts are at (–3, 0) and (3, 0).

The graph has no y-intercept. The two intercepts are sort of "endpoints" of the two parts of the curve. This curve is symmetric with respect to the y-axis. Figure 4-6 shows you the intercepts, a few other selected points, and the curves that are to the left and right of the intercepts.

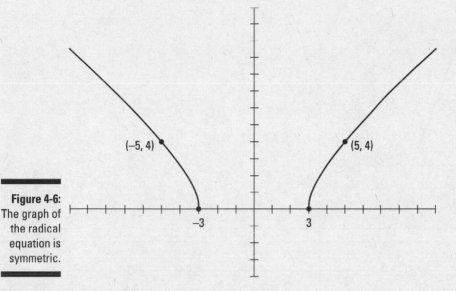

(−5, 4)(−5, 4)   (5, 4)

−3   3

Figure 4-6:
The graph of
the radical
equation is
symmetric.

Q. Sketch the graph of $y = 3|x + 4| - 2$.

A. The distinctive *V* shape in this graph opens upward because the number multiplying the absolute value operation is positive. The 3 multiplier also acts like a slope, making the curve relatively steep. The lowest *y*-value this curve can have is −2; that's because the term involving the absolute value can never be less than 0, and subtracting 2 lowers the *y*-value (0) by 2 units. The *V* shape, therefore, has its lowest point at (−4, −2). The −4 comes from the value that makes the absolute value term equal to 0.

The graph has a *y*-intercept of (0, 10). You get this value by letting $x = 0$ and solving for *y*. The graph also has two *x*-intercepts. You find them by letting $y = 0$ and then solving the absolute value equation $0 = 3|x + 4| - 2$.

This equation has two solutions, $x = -3\frac{1}{3}$, $x = -4\frac{2}{3}$. (Refer to Chapter 1 if you need info on solving absolute value equations.) Figure 4-7 shows the graph of this absolute value function.

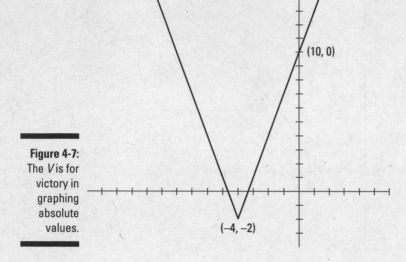

17. Sketch the graph of $y = \sqrt{4-x}$.

18. Sketch the graph of $y = \sqrt[3]{3x-1}$.

19. Sketch the graph of $y = |2x + 3| - 4$.

20. Sketch the graph of $y = 2|x - 3| + 1$.

Enter the Machines: Using a Graphing Calculator

Graphing calculators are wonderful instruments — and not just for graphing equations. Most graphing calculators allow you to perform intricate calculations, analyze data statistically, perform operations on matrices, and so on. The main drawback with graphing calculators, though, is that they give you what you ask for and not necessarily what you mean or want. When working with graphing calculators, it doesn't hurt to be too liberal with parentheses to be sure you've entered something into the calculator correctly (calculators strictly follow the order of operations — see Chapter 1).

You need a fairly good idea of what the graph should look like so you can adjust the viewing window to get a good picture of what's going on. If you expect to get a *U*-shaped curve and the graph you get is a line, then you know that either the viewing window is wrong or you put in the equation incorrectly. It's helpful to know the general shape, where the intercepts are, and how high and low the graph goes before you even start.

EXAMPLE

Q. Use a graphing calculator to graph $y = x(x - 13)^2(x + 11)$.

A. If you type this function into the graphing menu and use a "standard graph" command (it graphs from –10 to +10 both left and right and up and down), you don't see much of anything. Of course, you know that something should be there, because you know about intercepts and turning points. Looking at the factored equation, you see that the x-intercepts are at 0, 13, and –11. This graph is a fourth-degree polynomial, so you could have as many as three turning points (refer to "Graphing Basic Polynomial Curves" for details). So, for starters, the standard graphing window that goes left only to –10 isn't wide enough. Start out by changing the x-coordinates of your viewing window so that they go left to –13 (two less than the lowest x-intercept) and right to 15 (two larger than the highest x-intercept).

Then use the function on your calculator that automatically fits the curve to the x-width you've chosen to figure out how high and low the y-values should be. You get a graph that looks very much like the one in Figure 4-8. The curve touches the x-axis at $x = 13$, because the power on that factor is an even number. This even power causes a touch and go, a point at which the graph touches the axis but doesn't cross it (you have a double root). Odd powers on the factors allow the curve to cross sides of the x-axis.

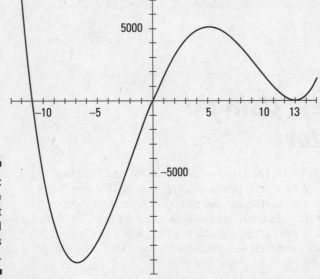

Figure 4-8:
The curve crosses at (–11, 0) and touches at 13.

Q. Use a graphing calculator to graph
$y = \dfrac{1}{x+2}$.

A. This perfectly swell rational function should graph very nicely on your calculator. The catch in graphing, though, is that you need to enter the equation correctly. If you enter "$y = 1 \div x + 2$" into your calculator, you get the wrong graph — the calculator follows the order of operations and divides before adding. Both the x and the 2 need to be in the denominator of the fraction, so put parentheses around them and enter "$y = 1 \div (x + 2)$."

The second issue you need to be aware of is that calculators tend to connect ends of the curves — even when they shouldn't be connected. The correct graph doesn't have connections between where x is smaller than –2 and where x is larger than –2 because at $x = -2$, the denominator is 0. (For more on what's happening in those gaps or holes in the graph, refer to Chapter 8.) You can turn your calculator to dot mode to prevent these stray connections — or just be aware that they shouldn't be there and ignore them. Figure 4-9 shows the graph of the equation. The window shows the graph from –5 to 5; that's wide enough.

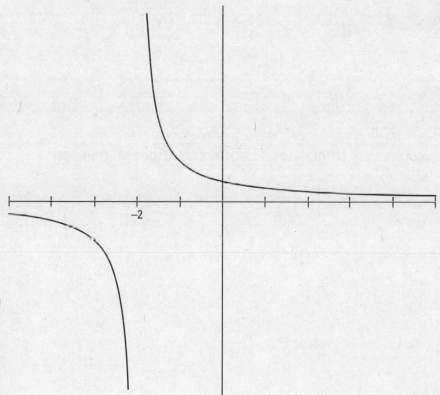

Figure 4-9:
The curve takes a break when $x = -2$.

21. Sketch the graph of $y = 3x + 12$.

22. Sketch the graph of $y = \sqrt{x^2 - 9}$.

23. Sketch the graph of $y = x^2(x^2 - 100)(x - 12)$.

24. Sketch the graph of $y = 2^{2x+1}$.

Answers to Problems on Graphing for the Good Life

The following are the answers to the practice problems presented in this chapter.

1 Name the points in Figure 4-2 that lie in Quadrant II or on the x-axis. The answer is **(–3, 4), (–7, 3), (–2, 1), (–6, 0), (5, 0).**

The points that lie in Quadrant II are in the upper-left corner of the coordinate plane. The x-coordinate of these points is always negative, and the y-coordinate is positive. Points that lie on the x-axis always have a y-coordinate of 0.

2 Name the points in Figure 4-2 that lie in Quadrant IV or on the y-axis. The answer is **(1, 2), (5, –3), (0, 4), (0, –3).**

The points that lie in Quadrant IV are in the lower-right corner of the coordinate plane. The y-coordinate of these points is always negative, and the x-coordinate is positive. Points that lie on the y-axis always have an x-coordinate of 0.

3 Identify the quadrant that contains all points with a negative x-coordinate and a negative y-coordinate. The answer is Quadrant III.

Both the x- and y-coordinates are negative for points in the third quadrant.

4 If you draw a segment connecting points (–4, 4) and (2, –2), what point lies on that line segment that has x- and y-coordinates with the same value (the numbers are the same)? The answer is **(0, 0).**

The segment connecting the two points contains points in which the coordinates have opposite values, such as –4 and 4, 3 and –3, –1 and 1, and so on. The only point on the segment that wouldn't have two coordinates with opposite signs is the origin — (0, 0).

5 Use intercepts and symmetry to sketch the graph of $y = x^2 - 9$.

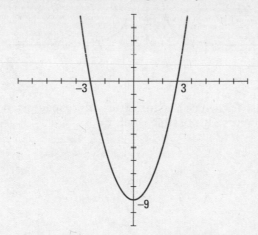

The intercepts are (0, –9), (3, 0), and (–3, 0). The curve is symmetric with respect to the y-axis.

6 Use intercepts and symmetry to sketch the graph of $y = 3 - |x|$.

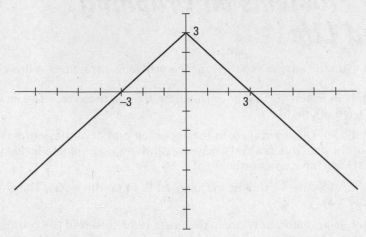

The intercepts are $(0, 3)$, $(3, 0)$, and $(-3, 0)$. The *V*-shaped absolute value curve is symmetric with respect to the *y*-axis.

7 Use intercepts and symmetry to sketch the graph of $y = \dfrac{1}{x^2}$.

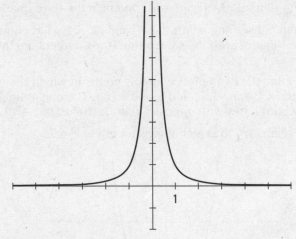

There are no intercepts. The curve is symmetric with respect to the *y*-axis. The *y*-values are always positive, and no curve is possible when $x = 0$.

8 Use intercepts and symmetry to sketch the graph of $y = \sqrt{16 - x^2}$.

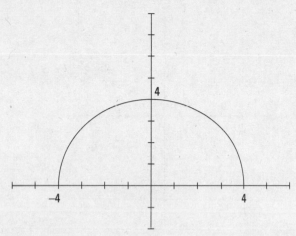

The intercepts are $(0, 4)$, $(4, 0)$, and $(-4, 0)$. The curve is symmetric with respect to the y-axis. The value of x can't be greater than 4 or less than -4, so the curve exists only between and including those two numbers.

9 Sketch the graph of $y = \frac{2}{3}x - 4$ using the slope and y-intercept.

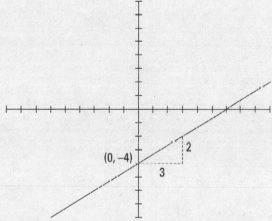

The line has a slope of $\frac{2}{3}$ and a y-intercept at $(0, -4)$. Mark a point for the y-intercept on your graph. Then count 3 units to the right and 2 units up from that y-intercept to locate another point on the line. Connect the two points to form the line.

10 Sketch the graph of $y = -5x + 1$ using the slope and y-intercept.

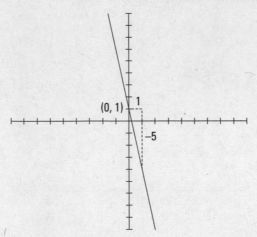

The line has a slope of –5 and a y-intercept at (0, 1). Mark a point for the y-intercept. Then count 1 unit to the right and 5 units down from that y-intercept to locate another point on the line. Connect the two points to form the line.

11 Sketch the graph of $2x + 3y = 6$ using the two intercepts; determine the slope of the line.

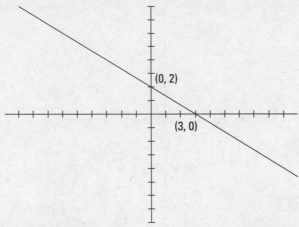

The slope is $-\dfrac{2}{3}$. The intercepts are (0, 2) and (3, 0). Draw a line through those two intercepts.

12 Sketch the graph of $9x - 2y = 18$ using the two intercepts; determine the slope of the line.

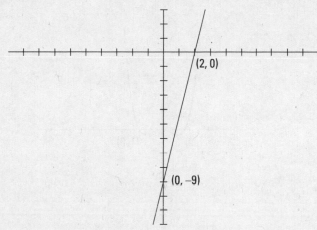

The slope is $\frac{9}{2}$. The intercepts are $(0, -9)$ and $(2, 0)$. Draw a line through the two intercepts.

13 Use intercepts, turning points, symmetry, and a few select points to sketch the graph of $y = x^3 - 4x^2$.

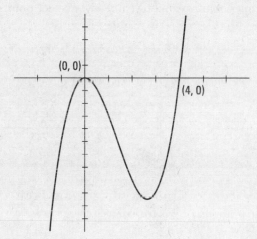

The intercepts are $(0, 0)$ and $(4, 0)$, and the graph has no symmetry. The curve just touches at the intercept $x = 0$, because, when factored as $y = x^2(x - 4)$, the power on the x-factor is an even number. The graph has two turning points.

14   Use intercepts, turning points, symmetry, and a few select points to sketch the graph of $y = x^4 - x^2$.

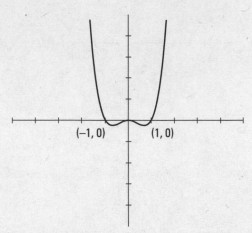

The intercepts are at $(0, 0)$, $(1, 0)$, and $(-1, 0)$. When factored, $y = x^2(x - 1)(x + 1)$. The curve is symmetric with respect to the y-axis. You have three turning points.

15   Use intercepts, turning points, symmetry, and a few select points to sketch the graph of $y = x(x + 3)(x - 4)(x - 6)$.

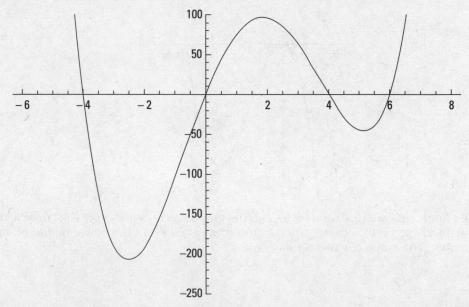

The intercepts are $(0, 0)$, $(-3, 0)$, $(4, 0)$, and $(6, 0)$. There's no symmetry. The graph has three turning points.

16 Use intercepts, turning points, symmetry, and a few select points to sketch the graph of
$y = (x^2 - 36)(x^2 - 25)$.

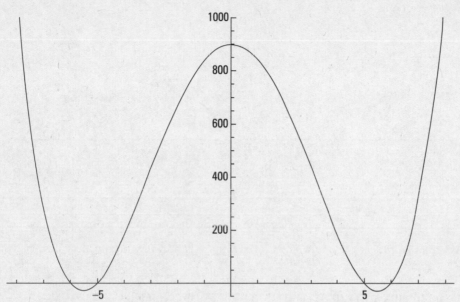

The intercepts are (6, 0), (–6, 0), (5, 0), and (–5, 0). The curve is symmetric with respect to the
y-axis. You can find three turning points.

17 Sketch the graph of $y = \sqrt{4 - x}$.

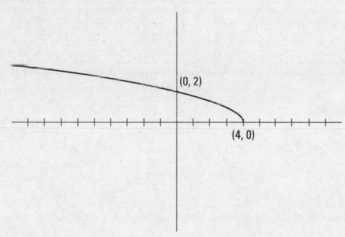

The intercepts are at (0, 2) and (4, 0). There is no graph when $x > 4$.

18 Sketch the graph of $y = \sqrt[3]{3x-1}$.

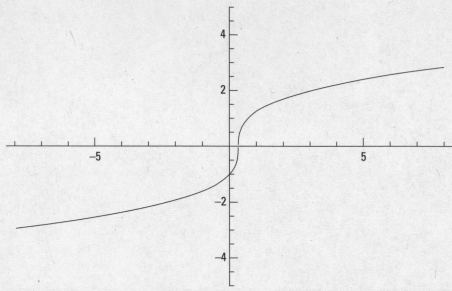

The intercepts are at $(0, -1)$ and $\left(\frac{1}{3}, 0\right)$. There's no symmetry.

19 Sketch the graph of $y = |2x + 3| - 4$.

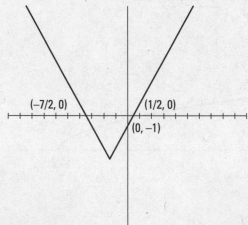

The intercepts are at $(0, -1)$, $\left(\frac{1}{2}, 0\right)$, and $\left(-\frac{7}{2}, 0\right)$.

20 Sketch the graph of $y = 2|x - 3| + 1$.

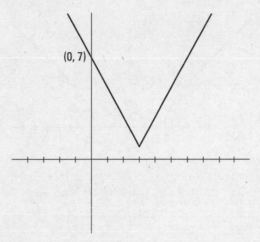

The only intercept is at (0, 7).

21 Sketch the graph of $y = 3x + 12$.

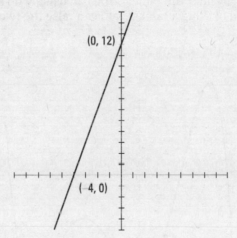

The intercepts arc at (0, 12) and (–4, 0). Set the window in the graphing screen to include a *y*-value that goes as high as the *y*-intercept, which is at 12.

22 Sketch the graph of $y = \sqrt{x^2 - 9}$.

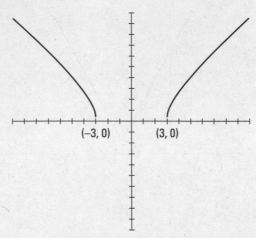

The intercepts are at $(-3, 0)$ and $(3, 0)$. You have no y-intercept. The curve is symmetric with respect to the y-axis. Be sure to enter the equation into the calculator very carefully, including both the x^2 term and -9 under the radical. You can usually do so by putting parentheses around the two terms. Instead of using a radical, you can raise the two terms in parentheses to the 0.5 power.

23 Sketch the graph of $y = x^2(x^2 - 100)(x - 12)$.

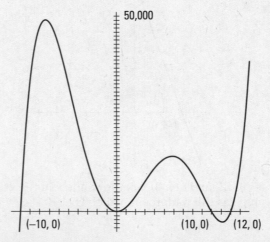

The intercepts are at $(0, 0)$, $(10, 0)$, $(-10, 0)$, and $(12, 0)$. Be sure to widen the graphing view to include all the x-intercepts. Then use the feature that automatically fits the window to graph the curve. You can see that the y-values go to a high of 50,000 and a low of $-5,000$.

24 Sketch the graph of $y = 2^{2x+1}$.

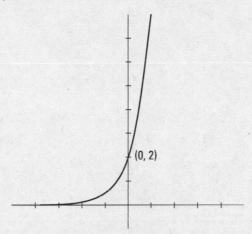

(0, 2)

The only intercept is at (0, 2). The x-axis is an asymptote, and the curve never crosses or touches it as it moves to the left. Be very careful when entering this equation into the calculator. You have to write the exponent, $2x + 1$, entirely in parentheses so the calculator can read it correctly.

Part II
Functions

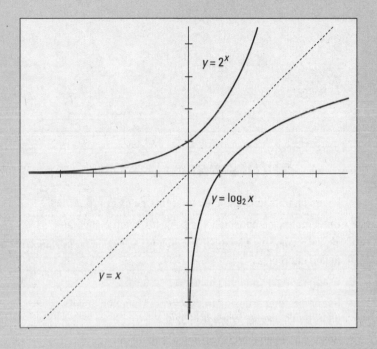

In this part . . .

- ✔ Focus on functions and their characteristics or properties.

- ✔ Use function composition and solve for function inverses.

- ✔ Tackle polynomial functions and their graphs using factorizations to find intercepts.

- ✔ Graph rational functions using asymptotes.

- ✔ Recognize properties of an exponential function and its logarithmic inverse when graphing.

- ✔ Solve exponential and logarithmic equations.

Chapter 5

Formulating Functions

● ●

In This Chapter

▶ Determining how much functions are worth

▶ Checking the restrictions on input and output: Domain and range

▶ Defining functions by their characteristics

▶ Performing operations on functions

▶ Looking at and creating inverse functions

● ●

A function is a very special type of relationship in mathematics. *Functions* are those relationships between input values and output values that guarantee that you'll never get more than one output for any input. For example, the equation $y = 8x + 2x^2 - 3^x$ is a function, because when you plug in any value for x, such as 2, you get only one value for y; in this case, $y = 15$. This characteristic of having just one value when you put a number into a function is essential when you're dealing with functions that model the cost of production or the number of amoebas in a culture. You don't want to be saying, "Well, there are either 16 or 16,000,000 amoebas in that dish." You'd prefer just one answer when you use the formula.

This chapter deals with how you can categorize a function and determine its characteristics. A function can be *even* or *odd* or even *piecewise*. Each label has its importance, and each has its place when studying and using functions.

Evaluating Functions

You read an equation like $y = 3x^3 - 4x^2 + 5x - 11$ with words such as "y is a function of x and is equal to three times x cubed minus four times . . . ." You get the point. From the equation, you determine that the y output is some function of the x input. You can plot the x's and y's in a graph. Another way of writing this function is $f(x) = 3x^3 - 4x^2 + 5x - 11$ (replacing the y with $f(x)$). This format is called *function notation*. The name of the function is f.

Using function notation when you need to list two or more functions is very handy, because you can name the functions f of x, g of x, h of x, and so on. Giving these different names helps to distinguish one function from another when you're working a bunch of them and they all do different things. When you use function notation, such as $f(x) = 3x^3 - 4x^2 + 5x - 11$, you *evaluate* the function for some x-value when you replace each x with the number in the parentheses and solve the equation. For example, using the rule $f(x)$ and evaluating for $x = 2$, you write that $f(2) = 3(2)^3 - 4(2)^2 + 5(2) - 11 = 3(8) - 4(4) + 10 - 11 = 24 - 16 - 1 = 7$.

Q. Find $f(2)$, $g(3)$, $h(-4)$, and $k(0)$ if $f(x) = 3x^3 - 4x^2 + 5x - 11$, $g(x) = |2x - 5| + 8$, $h(x) = \sqrt[3]{x + 31}$, and $k(x) = \dfrac{x^2 + 2}{x - 1}$.

A. Replace the x in each particular function with the number that's in the parentheses. Evaluate the expressions.

✔ $f(2) = 3(2)^3 - 4(2)^2 + 5(2) - 11 = 3(8) - 4(4) + 10 - 11 = 24 - 16 + 10 - 11 = 7$

✔ $g(3) = |2(3) - 5| + 8 = |6 - 5| + 8 = |1| + 8 = 1 + 8 = 9$

✔ $h(-4) = \sqrt[3]{(-4) + 31} = \sqrt[3]{27} = 3$

✔ $k(0) = \dfrac{(0)^2 + 2}{(0) - 1} = \dfrac{2}{-1} = -2$

---

1. If $f(x) = 3x^2 - 4x + 2$, find $f(-3)$.

2. If $g(x) = 5|2x - 3| - 1$, find $g(-3)$.

---

3. If $h(x) = \dfrac{12}{x^2 - 3}$, find $h(-3)$.

4. If $k(x) = \sqrt{25 - x^2}$, find $k(-3)$.

Determining the Domain and Range of a Function

The domain of a function is where it lives. No, that's not strictly true, but it does give you an idea of what the concept of domain involves. A function's *domain* consists of all the numbers that you can use as input values for that function. You always start out hoping or assuming that any real number is in the domain and can be used in the function's rule, but you have to discard numbers that cause inconsistencies or unreal numbers to appear when you perform the function's operations. If a function's rule involves a fraction, then you can't use any number in the function that would make the denominator of the fraction equal to 0. If a function's rule involves an even-powered radical, then you can't let the expression under the radical be a negative.

The *range* of a function consists of all the numbers you get when evaluating the input values. The *range* consists of all the real numbers that are results of performing the function operations on all the numbers in the domain.

Two ways of writing about domain and range are *inequality notation* and *interval notation* (see Chapter 1 for details on these formats). You may also write the numbers that aren't part of the answer set, like $x \neq -3, 3$.

Q. Determine the domain and range of the function $f(x) = \dfrac{24}{x^2 - 1}$.

A. **Domain: $x \neq -1, 1$; range: $f(x) \neq 0$.** The domain of $f(x)$ consists of all the real numbers except for $x = 1$ and $x = -1$. Both 1 and –1 make the denominator of the fraction equal to 0 — a situation that results in values that don't exist. You can't divide by 0. The way to describe this domain in interval notation is $(-\infty, -1) \cup (-1, 1) \cup (1, \infty)$ — the union symbol, $\cup$, means *or.* Interval notation is a bit cumbersome here, so I usually prefer just saying what's *not* allowed: $x \neq -1, 1$.

The range of $f(x)$ is all real numbers except for 0. When you input numbers into the rule, such as $f(2) = 8$, $f(-3) = 3$, $f\left(\dfrac{1}{2}\right) = -32$, and so on, you see, after a while, that you get big numbers and small numbers — both positive and negative — and the only number you can never get is 0. The numerator of a fraction has to equal 0 for the value of the fraction to be 0, so this function's rule will never result in 0. Saying that the range of $f(x)$ is all real numbers except 0, in interval notation, you write $(-\infty, 0) \cup (0, \infty)$. Otherwise, you can simply write that $f(x) \neq 0$.

Q. Determine the domain and range of the function $g(x) = \sqrt{6 - x}$.

A. **Domain: $x \leq 6$; range: $g(x) \geq 0$.** The domain of $g(x)$ is all real numbers smaller than or equal to 6, and the range of $g(x)$ is all positive numbers and 0 — domain: $(-\infty, 6]$; range: $[0, \infty)$. You can't have a negative number under a square root radical, so the input values can't be larger than 6. And when you take the square root of positive numbers or 0, you always get results that are either positive or 0.

---

5. Determine the domain and range of the function $f(x) = x^2 - 5$.

6. Determine the domain and range of the function $g(x) = |x + 7|$.

---

7. Determine the domain and range of the function $h(x) = \sqrt{x - 8}$.

8. Determine the domain and range of the function $k(x) = \dfrac{1}{x - 3}$.

Recognizing Even, Odd, and One-to-One Functions

The words *even, odd,* and *one-to-one* are ways of describing functions just like *perching, wading,* and *gull-like* are ways of describing birds. A perching bird still has many characteristics in common with other types of birds, but it has some unique characteristics that set it apart and make it special in a particular way.

Anyway, back to functions:

- **Even functions:** An even function has the same output regardless of whether you put in x or $-x$. Inputting a number and its opposite yields the same output value: $f(-x) = f(x)$. The graphs of even functions are symmetric with respect to the y-axis (see Chapter 4 for more on symmetry).

- **Odd functions:** An odd function does the opposite: The result of inputting x is the opposite of when you input $-x$ into the function rule: $f(-x) = -f(x)$. The graphs of odd functions are symmetric with respect to the origin (like a 180-degree turn).

- **One-to-one functions:** With a one-to-one function, not only do you get one output for every input value, but each output value also occurs exactly once. There's just one input value for every output value. If $f(x_1) = f(x_2)$, then $x_1 = x_2$.

Q. Describe the following functions as *even, odd,* or *neither;* then determine whether they're *one-to-one.*

a. $f(x) = \dfrac{4}{x^2 + 1}$

b. $g(x) = x + 2$

c. $h(x) = 4x - x^3$

A. Here are the answers:

a. Even, not one-to-one. Function f is an even function. For instance, you see that $f(1) = 2$ and $f(-1) = 2$. You get the same result. Substituting $-x$ into the function equation, you get $f(-x) = \dfrac{4}{(-x)^2 + 1} = \dfrac{4}{x^2 + 1}$. The function value is the same as for $f(x)$. The function f is not one-to-one, though, for the same reason that made it even. The output value 2 has two different input values; $f(1) = 2$ and $f(-1) = 2$. One-to-one functions have just one input value for every output value.

b. Neither, one-to-one. Function g is neither even nor odd. If you replace the x with $-x$, you get $g(-x) = (-x) + 2 = -x + 2$. This isn't the same as the original $g(x)$, and it isn't the opposite, either. To be the opposite, both terms would have to be negative. Function g is one-to-one, though. Every input value has just one output, and every output has just one input. You can confirm this conclusion by looking at the graph of g. It's a line that slants upward and doesn't double back on itself.

c. Odd, not one-to-one. Function h is an odd function. Replacing each x with $-x$, you get $h(-x) = 4(-x) - (-x)^3 = -4x + x^3 = -(4x - x^3)$. Function h isn't one-to-one, though. Both $h(2)$ and $h(-2)$ give you the same output — they're both equal to 0.

9. Determine whether the following function is *even, odd,* or *neither;* then determine whether it's *one-to-one:* $f(x) = x^2 - 5$.

10. Determine whether the following function is *even, odd,* or *neither;* then determine whether it's *one-to-one:* $g(x) = |x|$.

11. Determine whether the following function is *even, odd,* or *neither;* then determine whether it's *one-to-one:* $h(x) = x^3 + 1$.

12. Determine whether the following function is *even, odd,* or *neither;* then determine whether it's *one-to-one:* $k(x) = \sqrt[3]{x}$.

Composing Functions and Simplifying the Difference Quotient

You can add, subtract, multiply, and divide two or more functions by just taking their algebraic rules and performing those operations. For instance, if you want to add functions f and g, where $f(x) = 3x^2 - 9x + 7$ and $g(x) = -x^2 - 6x - 12$, then you just combine the like terms: $f + g = 3x^2 - 9x + 7 + (-x^2 - 6x - 12) = 3x^2 - x^2 - 9x - 6x + 7 - 12 = 2x^2 - 15x - 5$. Not much more to it than that.

Functions do have an operation, though, that's special to them — plain old numbers don't use this operation. The operation I'm referring to is composition. You *compose* functions by making one function the input of another. The operation symbol for composition is ∘, and you write it between the two functions you're composing, such as $f \circ g$.

The composition of functions f and g is defined as $f \circ g = f(g)$. This function is not commutative, so in general, $f \circ g \neq g \circ f$.

You can use the composition of functions when working with difference quotients. The difference quotient is the main component of the definition of the derivative in calculus. Most algebra books include exercises on simplifying the difference quotient to get you to practice a problem that you'll use when you get to calculus.

The *difference quotient* for the function $f(x)$ is $\dfrac{f(x+h) - f(x)}{h}$.

Q. Find both $f \circ g$ and $g \circ f$ if the functions $f(x) = 5x^2 - 2x + 1$ and $g(x) = 9 - 4x$.

A. $f \circ g = 388 - 352x + 80x^2$; $g \circ f = 5 + 8x - 20x^2$. First, $f \circ g = f(g) = f(9 - 4x)$. Replacing each x in the definition of function f with the rule for g, you get $f(9 - 4x) = 5(9 - 4x)^2 - 2(9 - 4x) + 1 = 5(81 - 72x + 16x^2) - 2(9 - 4x) + 1 = 405 - 360x + 80x^2 - 18 + 8x + 1 = 388 - 352x + 80x^2$. Now, reversing the roles, $g \circ f = g(5x^2 - 2x + 1) = 9 - 4(5x^2 - 2x + 1) = 9 - 20x^2 + 8x - 4 = 5 + 8x - 20x^2$. As you can see, by changing the order, you get different answers.

Q. Find the difference quotient for the function $f(x) = x^3 + 2x^2 - 3$.

A. $3x^2 + 3xh + h^2 + 4x + 2h$. You substitute $f(x + h)$ in for the first term in the numerator — you get the terms of $f(x + h)$ by replacing every x in the function's rule with an $x + h$. This step resembles composition if you think that $g(x) = x + h$ and that you're composing $f(g)$. Then substitute the original function, $f(x)$, in for the second term. Be sure to put the values defining $f(x)$ in parentheses so that the negative sign gets distributed over each term:

$$\frac{f(x+h) - f(x)}{h} = \frac{(x+h)^3 + 2(x+h)^2 - 3 - (x^3 + 2x^2 - 3)}{h}$$

$$= \frac{x^3 + 3x^2h + 3xh^2 + h^3 + 2(x^2 + 2xh + h^2) - 3 - x^3 - 2x^2 + 3}{h}$$

$$= \frac{x^3 + 3x^2h + 3xh^2 + h^3 + 2x^2 + 4xh + 2h^2 - 3 - x^3 - 2x^2 + 3}{h}$$

After cubing and squaring the binomial and distributing over the parentheses, you find pairs of terms and their opposites in the numerator. You can eliminate them, and then you can factor out an h and reduce the fraction:

$$\frac{\cancel{x^3} + 3x^2h + 3xh^2 + h^3 + \cancel{2x^2} + 4xh + 2h^2 \cancel{-3} \cancel{-x^3} \cancel{-2x^2} + \cancel{3}}{h}$$

$$= \frac{3x^2h + 3xh^2 + h^3 + 4xh + 2h^2}{h} = \frac{\cancel{h}(3x^2 + 3xh + h^2 + 4x + 2h)}{\cancel{h}}$$

$$= 3x^2 + 3xh + h^2 + 4x + 2h$$

This answer probably doesn't seem like much, but it's one step short of finding the derivative. An important part of the result is that you get rid of the denominator with the h in it.

---

13. Find both $f \circ g$ and $g \circ f$ if $f(x) = 2x^2 + 3x - 4$ and $g(x) = x - 2$.

14. Find both $f \circ g$ and $g \circ f$ if $f(x) = 2x + 7$ and $g(x) = 5 - 3x$.

15. Find both $f \circ g$ and $g \circ f$ if $f(x) = x^2 - 9$ and $g(x) = x^3 + 3$.

16. Find both $f \circ g$ and $g \circ f$ if $f(x) = f(x) = 2x - 8$ and $g(x) = \frac{x+8}{2}$.

17. Evaluate the difference quotient for $f(x) = 3x^2 + 2x - 5$.

18. Evaluate the difference quotient for $f(x) = \frac{3x+1}{x-2}$.

19. Evaluate the difference quotient for $f(x) = \sqrt{x-3}$.

20. Evaluate the difference quotient for $f(x) = 4$.

Solving for Inverse Functions

An *inverse operation* does the opposite of what the original operation does. For instance, the inverse of adding 6 to a number is subtracting 6. The inverse of dividing by 4 is multiplying by 4. An *inverse function* undoes or performs the opposite of what the original function did.

Consider the function $f(x) = \frac{3x+2}{x-4}$. Letting $x = 2$, you get $f(2) = -4$:

$f(2) = \frac{3(2)+2}{(2)-4} = \frac{6+2}{-2} = \frac{8}{-2} = -4$. The inverse function for f, designated by

f^{-1}, is $f^{-1}(x) = \dfrac{4x+2}{x-3}$. Look what happens when I put -4 in for x in the inverse:

$f^{-1}(-4) = \dfrac{4(-4)+2}{(-4)-3} = \dfrac{-16+2}{-7} = \dfrac{-14}{-7} = 2$. The result is 2, the number that I initially put in

function f. The inverse gets you back to the original input.

You may wonder where I found the inverse for the function f. I used the following process. But first, be aware that only one-to-one functions have inverses. When you put a result into an inverse function, it's important that you get just one answer. And that's the case only with one-to-one functions. Go to "Recognizing Even, Odd, and One-to-One Functions," earlier in this chapter, if you need more information.

To solve for an inverse function:

1. **Change $f(x)$ to y for convenience.**

2. **Switch all the x's in the function equation to y's and the y to an x.**

3. **Solve for y in this new equation.**

4. **Replace the y with $f^{-1}(x)$.**

Q. Find the inverse for $f(x) = \dfrac{3x+2}{x-4}$.

A. Using the four steps, first replace the $f(x)$ with y to get $y = \dfrac{3x+2}{x-4}$. Next, switch all the x's and y to get $x = \dfrac{3y+2}{y-4}$. To solve for y, first multiply each side of the equation by the denominator, $y - 4$. Then distribute the x over the binomial. Next, add $4x$ to each side and subtract $3y$ from each side to get all the terms with y in them on the left side of the equation:

$$x(y-4) = 3y+2$$
$$xy - 4x = 3y + 2$$
$$xy - 3y = 4x + 2$$

Now factor out the y on the left and divide each side of the equation by $x - 3$. Change the y to $f^{-1}(x)$:

$$y(x-3) = 4x+2$$
$$y = \dfrac{4x+2}{x-3}$$
$$f^{-1}(x) = \dfrac{4x+2}{x=3}$$

(Yes, this is the inverse from the example in the main text. Thought you should see where it came from!)

Note: The composition of functions is usually *not* commutative, but when you compose a function and its inverse, the result of both compositions is x. So, with this special combination, you do have commutativity.

21. Find the inverse function for $f(x) = 3x - 1$.

22. Find the inverse function for $f(x) = x^3 + 3$.

23. Find the inverse function for $f(x) = \frac{x-1}{2x+1}$.

24. Find the inverse function for $f(x) = \frac{2x-2}{3x-2}$.

Answers to Problems on Formulating Functions

The following are the answers to the practice problems presented earlier in this chapter.

1 If $f(x) = 3x^2 - 4x + 2$, find $f(-3)$. The answer is **$f(-3) = 41$.**

Replace all the x's in the function equation with -3 and simplify: $f(-3) = 3(-3)^2 - 4(-3) + 2 = 3(9) + 12 + 2 = 41$.

2 If $g(x) = 5|2x - 3| - 1$, find $g(-3)$. The answer is **$g(-3) = 44$.**

Replace the x inside the absolute value operation with the -3. Simplify what's in the absolute value operation, and then find the absolute value before multiplying by 5 and subtracting 1:

$$g(-3) = 5|2(-3) - 3| - 1 = 5|-9| - 1 = 5(9) - 1 = 45 - 1 = 44$$

3 If $h(x) = \dfrac{12}{x^2 - 3}$, find $h(-3)$. The answer is **$h(-3) = 2$.**

Replace the x in the denominator with -3. After squaring the -3 and subtracting 3, divide the numerator by the result: $\dfrac{12}{(-3)^2 - 3} = \dfrac{12}{9 - 3} = \dfrac{12}{6}$.

4 If $k(x) = \sqrt{25 - x^2}$, find $k(-3)$. The answer is **$k(-3) = 4$.**

Subtracting $(-3)^2 = 9$ from 25 under the radical, find the square root of 16, which is 4.

5 Determine the domain and range of the function $f(x) = x^2 - 5$. The answer is **domain: all real numbers, or $(-\infty, \infty)$; range: $y = f(x) \geq -5$, or $[-5, \infty)$.**

Substitute in any real number for x, and the result is always a real number, so the domain is all reals. The lowest number that's an output value is -5. That's because when you square x, it becomes positive. The smallest x^2 can be is 0, and when you subtract 5, you get -5. All the other values in the range are larger than that, and they go up without bound.

6 Determine the domain and range of the function $g(x) = |x + 7|$. The answer is **domain: all real numbers, or $(-\infty, \infty)$; range: $y = g(x) \geq 0$, or $[0, \infty)$.**

You can substitute in any real number for x, and the result will be a real number, so the domain is all reals. The lowest number that's an output value is 0, because the absolute value of any number is positive — except for 0, which is as low as the range can go.

7 Determine the domain and range of the function $h(x) = \sqrt{x - 8}$. The answer is **domain: $x \geq 8$ or $[8, \infty)$; range: $y = h(x) \geq 0$, or $[0, \infty)$.**

You can't have a negative under the square root radical, so the smallest number that you can substitute in for x is 8. The results of taking a square root are always positive or at least 0.

8 Determine the domain and range of the function $k(x) = \dfrac{1}{x - 3}$. The answer is **domain: $x \neq 3$, or $(-\infty, 3) \cup (3, \infty)$; range: $y = k(x) \neq 0$, or $(-\infty, 0) \cup (0, \infty)$.**

You can't have a 0 in the denominator, so x can't be 3. The function output is any real number except 0, because a fraction can equal 0 only if the numerator is equal to 0. The numerator is always the number 1.

9 Determine whether the function is *even*, *odd*, or *neither*; then determine whether it's *one-to-one*: $f(x) = x^2 - 5$. The function is **even; not one-to-one.**

The function is an even function, because when you replace the x's with $-x$'s, you get the same function definition: $f(-x) = (-x)^2 - 5 = x^2 - 5 = f(x)$. The function isn't one-to-one, because you find more than one x-value for all but one y-value. For instance, if $f(x) = 11$, x can be either 4 or -4.

10 Determine whether the function is *even, odd,* or *neither;* then determine whether it's one-to-one: $g(x) = |x|$. The function is **even; not one-to-one.**

The function is an even function, because when you replace the x's with $-x$'s, you get the same function definition: $g(-x) = |-x| = |x| = g(x)$. The function isn't one-to-one because it has more than one x-value for all but one y-value. For instance, if $g(x) = 3$, x can be either 3 or -3.

11 Determine whether the function is *even, odd,* or *neither;* then determine whether it's one-to-one: $h(x) = x^3 + 1$. The function is **neither even nor odd; one-to-one.**

The function isn't even or odd, because when you replace the x's with $-x$'s, you don't get the same function definition and you don't get the opposite: $h(-x) = (-x)^3 + 1 = -x^3 + 1 \neq x^3 + 1$. Also, $-x^3 + 1 \neq -(x^3 + 1)$. The function is one-to-one because every y-value has only one x-value and every x-value has only one y-value.

12 Determine whether the function is *even, odd,* or *neither;* then determine whether it's one-to-one: $k(x) = \sqrt[5]{x}$. The function is **odd; one-to-one.**

The function is an odd function, because when you replace the x's with $-x$'s, you get the opposite of the function definition: $k(-x) = \sqrt[5]{-x} = -\sqrt[5]{x}$. It's also one-to-one, because it's a fifth root, and when you take the fifth root of a number, its sign doesn't change.

13 Find both $f \circ g$ and $g \circ f$ if $f(x) = 2x^2 + 3x - 4$ and $g(x) = x - 2$. The answer is $f \circ g = 2x^2 - 5x - 2$; $g \circ f = 2x^2 + 3x - 6$.

First, to compute the composition $f \circ g$, you follow these steps:

$$
\begin{aligned}
f \circ g &= 2(x-2)^2 + 3(x-2) - 4 \\
&= 2(x^2 - 4x + 4) + 3(x-2) - 4 \\
&= 2x^2 - 8x + 8 + 3x - 6 - 4 \\
&= 2x^2 - 5x - 2
\end{aligned}
$$

Next, the composition $g \circ f$ is

$$
\begin{aligned}
g \circ f &= (2x^2 + 3x - 4) - 2 \\
&= 2x^2 + 3x - 6
\end{aligned}
$$

14 Find both $f \circ g$ and $g \circ f$ if $f(x) = 2x + 7$ and $g(x) = 5 - 3x$. The answer is $f \circ g = 17 - 6x$; $g \circ f = -16 - 6x$.

First compute the composition $f \circ g$:

$$
\begin{aligned}
f \circ g &= 2(5 - 3x) + 7 \\
&= 10 - 6x + 7 \\
&= 17 - 6x
\end{aligned}
$$

Then, the composition $g \circ f$ is

$$
\begin{aligned}
g \circ f &= 5 - 3(2x + 7) \\
&= 5 - 6x - 21 \\
&= -16 - 6x
\end{aligned}
$$

15 Find both $f \circ g$ and $g \circ f$ if $f(x) = x^2 - 9$ and $g(x) = x^3 + 3$. The answer is $f \circ g = x^6 + 6x^3$; $g \circ f = x^6 - 27x^4 + 243x^2 - 726$.

The composition $f \circ g$ is

$$
\begin{aligned}
f \circ g &= (x^3 + 3)^2 - 9 \\
&= x^6 + 6x^3 + 9 - 9 \\
&= x^6 + 6x^3
\end{aligned}
$$

Here are the steps to find $g \circ f$:

$$
\begin{aligned}
g \circ f &= (x^2 - 9)^3 + 3 \\
&= x^6 - 27x^4 + 243x^2 - 729 + 3 \\
&= x^6 - 27x^4 + 243x^2 - 726
\end{aligned}
$$

16 Find both $f \circ g$ and $g \circ f$ if $f(x) = f(x) = 2x - 8$ and $g(x) = \frac{x+8}{2}$. The answer is $f \circ g = x$; $g \circ f = x$.

First compute $f \circ g$: $f \circ g = 2\left(\frac{x+8}{2}\right) - 8 = x + 8 - 8 = x$. The composition $g \circ f$ is

$g \circ f = \frac{(2x-8)+8}{2} = \frac{2x}{2} = x$. In this case, the composition is commutative; the answers are the same.

You get this result when the two functions are inverses of one another.

17 Evaluate the difference quotient for $f(x) = 3x^2 + 2x - 5$. The answer is $6x + 3h + 2$.

Compute the difference quotient:

$$
\begin{aligned}
\frac{f(x+h) - f(x)}{h} &= \frac{3(x+h)^2 + 2(x+h) - 5 - (3x^2 + 2x - 5)}{h} \\
&= \frac{3(x^2 + 2xh + h^2) + 2(x+h) - 5 - (3x^2 + 2x - 5)}{h} \\
&= \frac{3x^2 + 6xh + 3h^2 + 2x + 2h - 5 - 3x^2 - 2x + 5}{h} \\
&= \frac{\cancel{3x^2} + 6xh + 3h^2 + \cancel{2x} + 2h \cancel{-5} \cancel{-3x^2} \cancel{-2x} \cancel{+5}}{h} \\
&= \frac{6xh + 3h^2 + 2h}{h} = \frac{\cancel{h}(6x + 3h + 2)}{\cancel{h}} = 6x + 3h + 2
\end{aligned}
$$

18 Evaluate the difference quotient for $f(x) = \frac{3x+1}{x-2}$. The answer is $\frac{-7}{(x+h-2)(x-2)}$.

Compute the difference quotient, $\dfrac{f(x+h) - f(x)}{h} = \dfrac{\dfrac{3(x+h)+1}{(x+h)-2} - \dfrac{3x+1}{x-2}}{h}$. You have a complex fraction — a fraction within a fraction. The best approach here is to find a common denominator for the two fractions in the numerator, write them each with that common denominator, and combine them; then multiply the numerator times the reciprocal of the denominator:

$$\frac{\frac{3(x+h)+1}{(x+h)-2} \cdot \frac{x-2}{x-2} - \frac{3x+1}{x-2} \cdot \frac{(x+h)-2}{(x+h)-2}}{h}$$

$$= \frac{\frac{(3x+3h+1)(x-2)}{(x+h-2)(x-2)} - \frac{(3x+1)(x+h-2)}{(x-2)(x+h-2)}}{h}$$

$$= \frac{\frac{3x^2-6x+3xh-6h+x-2}{(x+h-2)(x-2)} - \frac{3x^2+3xh-6x+x+h-2}{(x-2)(x+h-2)}}{h}$$

$$= \frac{\frac{3x^2-6x+3xh-6h+x-2-3x^2-3xh+6x-x-h+2}{(x+h-2)(x-2)}}{h}$$

$$= \frac{-7h}{(x+h-2)(x-2)} \cdot \frac{1}{h} = \frac{-7}{(x+h-2)(x-2)}$$

19 Evaluate the difference quotient for $f(x) = \sqrt{x-3}$. The answer is: $\dfrac{1}{\sqrt{x+h-3}+\sqrt{x-3}}$.

Compute the difference quotient: $\dfrac{f(x+h)-f(x)}{h} = \dfrac{\sqrt{x+h-3}-\sqrt{x-3}}{h}$.

You have two radicals in the numerator that don't combine. Multiply both numerator and denominator by the conjugate of the numerator. The product in the numerator consists of two binomials where you have the difference and sum of the same terms. Multiply and simplify.

$$\frac{\sqrt{x+h-3}-\sqrt{x-3}}{h} \cdot \frac{\sqrt{x+h-3}+\sqrt{x-3}}{\sqrt{x+h-3}+\sqrt{x-3}}$$

$$= \frac{\left(\sqrt{x+h-3}-\sqrt{x-3}\right)\left(\sqrt{x+h-3}+\sqrt{x-3}\right)}{h\left(\sqrt{x+h-3}+\sqrt{x-3}\right)}$$

$$= \frac{\left(\sqrt{x+h-3}\right)^2 - \left(\sqrt{x-3}\right)^2}{h\left(\sqrt{x+h-3}+\sqrt{x-3}\right)}$$

$$= \frac{x+h-3-(x-3)}{h\left(\sqrt{x+h-3}+\sqrt{x-3}\right)} = \frac{h}{h\left(\sqrt{x+h-3}+\sqrt{x-3}\right)}$$

$$= \frac{h}{h\left(\sqrt{x+h-3}+\sqrt{x-3}\right)} = \frac{1}{\sqrt{x+h-3}+\sqrt{x-3}}$$

20 Evaluate the difference quotient for $f(x) = 4$. The answer is **0**.

Compute the difference quotient; this problem is easy and hard at the same time. You have no variable to insert the input values into: $\dfrac{f(x+h)-f(x)}{h} = \dfrac{4-4}{h}$.

Simplifying, you get $\dfrac{0}{h} = 0$.

21 Find the inverse function for $f(x) = 3x - 1$. The answer is $f^{-1}(x) = \dfrac{x+1}{3}$.

Use the steps to solve for the inverse function:

$$y = 3x - 1$$
$$x = 3y - 1$$
$$x + 1 = 3y$$
$$\frac{x+1}{3} = y$$

22 Find the inverse function for $f(x) = x^3 + 3$. The answer is $f^{-1}(x) = \sqrt[3]{x-3}$.

Use the steps to solve for the inverse function:

$$y = x^3 + 3$$
$$x = y^3 + 3$$
$$x - 3 = y^3$$
$$\sqrt[3]{x-3} = \sqrt[3]{y^3}$$
$$\sqrt[3]{x-3} = y$$

23 Find the inverse function for $f(x) = \dfrac{x-1}{2x+1}$. The answer is $f^{-1}(x) = \dfrac{x+1}{1-2x}$.

Using the steps to solve for the inverse function:

$$y = \frac{x-1}{2x+1}$$
$$x = \frac{y-1}{2y+1}$$
$$x(2y+1) = y - 1$$
$$2xy + x = y - 1$$
$$2xy - y = -x - 1$$
$$y(2x-1) = -x - 1$$
$$y = \frac{-x-1}{2x-1} = \frac{x+1}{1-2x}$$

In the last step, I multiplied both numerator and denominator by -1 to reverse all the signs.

24 Find the inverse function for $f(x) = \dfrac{2x-2}{3x-2}$. The answer is $f^{-1}(x) = \dfrac{2x-2}{3x-2}$.

Notice that this function is its own inverse! Sorta neat, don't you think? You get this result when the graph of a function is symmetric with respect to the line $y = x$. To find this answer, use the steps to solve for the inverse function:

$$y = \frac{2x-2}{3x-2}$$
$$x = \frac{2y-2}{3y-2}$$
$$x(3y-2) = 2y - 2$$
$$3xy - 2x = 2y - 2$$
$$3xy - 2y = 2x - 2$$
$$y(3x-2) = 2x - 2$$
$$y = \frac{2x-2}{3x-2}$$

Chapter 6

Specializing in Quadratic Functions

· ·

In This Chapter

▶ Picking up on the characteristics of a quadratic from its equation

▶ Using quadratic functions to solve real-life problems

▶ Getting the visual: Parabolic graphs

· ·

Quadratic functions are relatively recognizable, easily manipulated, and surprisingly common in the realm of the functions that model what goes on in the real world — perfect pawns for all your brilliant plots (in algebra)!

The standard form for the quadratic function is $f(x) = ax^2 + bx + c$ where a, b, and c are real numbers that are constants, $a \neq 0$, and x is the variable, or input value.

From this simple format, you can tell a lot about the nature of the function. Solving for intercepts or other function values is possible because of the factoring properties of the quadratics and the quadratic formula (see Chapter 2 for more info on solving quadratic equations). Saying anything bad about these functions is just really hard — so I won't!

In this chapter, I show you how to find the x- and y-intercepts of a quadratic function. Every quadratic function has a y-intercept, but not all of them have x-intercepts — here, you can find out how to recognize which case you have. You can also see what wonderful practical models quadratic functions can be in the real world. Finally, I tell you how to graph quadratic functions to get those smooth, U-shaped curves.

Finding Intercepts and the Vertex of a Parabola

The *intercepts* of a function are where the curve or graph of the function crosses an axis. The x-intercepts of a function always have a 0 for the y-coordinate. They look like $(-3, 0)$, $(4, 0)$, and so on. A function has only one y-intercept (because of the definition of a function), and the y-intercept always has a 0 for its x-coordinate, such as $(0, 6)$.

To solve for the x-intercepts of a quadratic function, you set $f(x)$ equal to 0 (this is the same as setting y equal to 0) and then solve for x. To solve for the y-intercept of a quadratic function, you set x equal to 0 and evaluate $f(0)$.

The graph of a quadratic function is a parabola, a *U*-shaped curve; the parabola's *vertex* is a point that's at the highest or lowest position of the graph of the function. The vertex is a parabola's only turning point. Here are a couple things to note:

- ✔ When the constant *a* is positive, the parabola opens upward and the vertex is in the lowest position (it's the *minimum*).

- ✔ When *a* is negative, the parabola opens downward, and the vertex is in the highest position (the *maximum*).

You can find the *x*-coordinate of the vertex using $x = -\dfrac{b}{2a}$. Calculate the *x*-coordinate, and then substitute that number into the equation of the function to find the *y*-coordinate of the vertex.

Q. Find the intercepts and vertex of the function $f(x) = 4x^2 + 8x - 21$.

A. **y-intercept: (0, –21); x-intercepts:** $\left(-\dfrac{7}{2}, 0\right), \left(\dfrac{3}{2}, 0\right)$**; vertex: (–1, –25).** To find the *y*-intercept, let all the *x*'s equal 0 in the equation; you get $y = 4(0)^2 + 8(0) - 21 = -21$, so the *y*-intercept is (0, –21). For the *x*-intercepts, let $f(x) = 0$ and solve $0 = 4x^2 + 8x - 21$. Factoring, $0 = (2x + 7)(2x - 3)$, which has the solutions $x = -\dfrac{7}{2}, \dfrac{3}{2}$. The

x-intercepts are $\left(-\dfrac{7}{2}, 0\right), \left(\dfrac{3}{2}, 0\right)$.

You can find the *x*-coordinate of the vertex using $a = 4$ and $b = 8$ in the equation, and get $x = -\dfrac{8}{2(4)} = -1$. Substitute the –1 in for the *x*'s to solve for the *y*-coordinate of the vertex: $y = 4(-1)^2 + 8(-1) - 21 = 4 - 8 - 21 = -25$. The coordinates of the vertex are (–1, –25).

---

1. Find the intercepts and vertex of the function $f(x) = 2x^2 - 9x - 18$.

2. Find the intercepts and vertex of the function $f(x) = 25 - 4x^2$.

---

3. Find the intercepts and vertex of the function $f(x) = x^2 + 10x + 7$.

4. Find the intercepts and vertex of the function $f(x) = -x^2 - 8x - 16$.

Applying Quadratics to Real-Life Situations

Quadratic equations lend themselves to modeling situations that happen in real life, such as the rise and fall of profits from selling goods, the decrease and increase in the amount of time it takes to run a mile based on your age, and so on. The wonderful part of having something that can be modeled by a quadratic is that you can easily solve the equation when set equal to zero and predict the patterns in the function values.

The vertex and *x*-intercepts are especially useful. These intercepts tell you where numbers change from positive to negative or negative to positive, so you know, for instance, where the ground is located in a physics problem or when you'd start making a profit or losing money in a business venture. The vertex tells you where you can find the absolute maximum or minimum cost, profit, speed, height, time, or whatever you're modeling.

Q. In 1972, you could buy a Mercury Comet for about $3,200. (Mine had bucket seats and was really spiffy.) Cars can depreciate in value pretty quickly, but a 1972 Comet in pristine condition may be worth a lot of money to a collector today. Let the value of one of these Comets be modeled by the quadratic function $v(t) = 18.75t^2 - 450t + 3,200$, where *t* is the number of years since 1972. When is the value of the function equal to 0 (what is an *x*-intercept), what was the car's lowest value, and what was its value in 2010?

A. **The car's value never dropped to 0, the lowest value was $500, and the car was worth $13,175 in the year 2010.** In this model, the *y*-intercept represents the initial value. When $t = 0$, the function is $v(0) = 3,200$, which corresponds to the purchase price. Find the *x* intercepts by solving $18.75t^2 - 450t + 3,200 = 0$. Using the quadratic formula (you could try factoring, but it's a bit of a challenge and, as it turns out, the equation doesn't factor), you get $-37,500$ under the radical in the formula. You can't get a real-number solution, so the graph has no *x*-intercept. The value of the Comet doesn't ever get down to 0. Find the lowest value by determining the vertex. Using the formula, $t = \frac{-(-450)}{2(18.75)} = \frac{450}{37.50} = 12$. This coordinate tells you that 12 years from the beginning (1984 — add 12 to 1972), the value of the Comet is at its lowest. Replace the *t*'s in the formula with 12, and you get $v(12) = 18.75(12)^2 - 450(12) + 3,200 = 500$. The Comet was worth $500 in 1984. To find the value of the car in 2010, you let $t = 38$, because the year 2010 is 38 years after 1972. The value of the car in 2010 is $v(38) = 18.75(38)^2 - 450(38) + 3,200 = \$13,175$. (Makes me wish I had kept my Comet!)

5. The height of a ball *t* seconds after it's thrown into the air from the top of a building can be modeled by $h(t) = -16t^2 + 48t + 64$, where $h(t)$ is height in feet. How high is the building, how high does the ball rise before starting to drop downward, and after how many seconds does the ball hit the ground?

6. The profit function telling Georgio how much money he will net for producing and selling *x* specialty umbrellas is given by $P(x) = -0.00405x^2 + 8.15x - 100$. What is Georgio's loss if he doesn't sell any of the umbrellas he produces, how many umbrellas does he have to sell to break even, and how many does he have to sell to earn the greatest possible profit?

7. Chip ran through a maze in less than a minute the first time he tried. His times got better for a while with each new try, but then his times got worse (he took longer) due to fatigue. The amount of time Chip took to run through the maze on the ath try can be modeled by $T(a) = 0.5a^2 - 9a + 48.5$. How long did Chip take to run the maze the first time, and what was his best time?

8. A highway underpass is parabolic in shape. If the curve of the underpass can be modeled by $h(x) = 50 - 0.02x^2$, where x and $h(x)$ are in feet, then how high is the highest point of the underpass, and how wide is it?

Graphing Parabolas

The graph of a quadratic function is a smooth, U-shaped curve that opens either upward or downward, depending on the sign of the coefficient of the x^2 term. The vertex and intercepts offer the quickest, easiest points to help with the graph of the parabola. You can resort to solving for other points if the graph has no x-intercepts or if you need additional information to determine more about the shape.

Another aid to use when graphing parabolas is the *axis of symmetry;* a parabola is symmetric about a vertical line that runs through the vertex. Points on either side of the axis of symmetry that have the same y-value are equal distances from the axis. The equation of the *axis of symmetry* is $x = h$, where (h, k) is the vertex of the parabola.

EXAMPLE

Q. Sketch the graph of the parabola $f(x) = -x^2 + 6x + 40$, labeling any intercepts and the vertex and showing the axis of symmetry.

A. As you can see in Figure 6-1, the y-intercept is $(0, 40)$; you can find it by letting all the x's equal 0 and simplifying. Find the x-intercepts by setting $-x^2 + 6x + 40$ equal to 0 and factoring: $0 = -(x^2 - 6x - 40) = -(x + 4)(x - 10)$; $x = -4$ and 10, so the intercepts are at $(-4, 0)$ and $(10, 0)$. The vertex is at $(3, 49)$: You find the x-value using the formula (in the earlier section "Finding Intercepts and the Vertex of a Parabola") and then replace the x's with 3s and simplify for the y-coordinate.

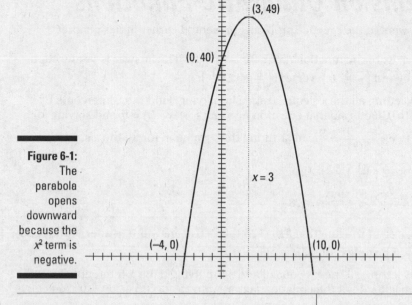

Figure 6-1:
The parabola opens downward because the x^2 term is negative.

(3, 49)

(0, 40)

$x = 3$

(−4, 0)

(10, 0)

---

9. Sketch the graph of the parabola $f(x) = 4x^2$, labeling any intercepts and the vertex and showing the axis of symmetry.

10. Sketch the graph of the parabola $f(x) = -\frac{1}{3}x^2 + 3$, labeling any intercepts and the vertex and showing the axis of symmetry.

11. Sketch the graph of the parabola $f(x) = 3x^2 - 6x - 9$, labeling any intercepts and the vertex and showing the axis of symmetry.

12. Sketch the graph of the parabola $f(x) = -2x^2 + 10x - 8$, labeling any intercepts and the vertex and showing the axis of symmetry.

13. Sketch the graph of the parabola $f(x) = \frac{1}{2}x^2 + 3x + 1$, labeling any intercepts and the vertex and showing the axis of symmetry.

14. Sketch the graph of the parabola $f(x) = -x^2 + 6x$, labeling any intercepts and the vertex and showing the axis of symmetry.

Answers to Problems on Quadratic Functions

The following are the answers to the practice problems presented earlier in this chapter.

1 Find the intercepts and vertex of the function $f(x) = 2x^2 - 9x - 18$. The answer is **y-intercept: (0, –18); x-intercepts: (6, 0) and $\left(-\dfrac{3}{2}, 0\right)$; vertex: $\left(\dfrac{9}{4}, -28\dfrac{1}{8}\right)$.**

You find the y-intercept by letting all the x's equal 0 and simplifying. Find the x-intercepts by setting $2x^2 - 9x - 18$ equal to 0, factoring that equation into $(2x + 3)(x - 6) = 0$, and solving for x.

The x-value of the vertex is $x = \dfrac{-(-9)}{2(2)} = \dfrac{9}{4}$. Substituting the x-value in for all the x's, you get

$$f\left(\frac{9}{4}\right) = 2\left(\frac{9}{4}\right)^2 - 9\left(\frac{9}{4}\right) - 18 = 2\left(\frac{81}{16}\right) - \frac{81}{4} - 18$$

$$= \frac{162}{16} - \frac{324}{16} - \frac{288}{16} = \frac{-450}{16} = -\frac{225}{8} = -28\frac{1}{8}$$

2 Find the intercepts and vertex of the function $f(x) = 25 - 4x^2$. The answer is **y-intercept: (0, 25); x-intercepts: $\left(\dfrac{5}{2}, 0\right), \left(-\dfrac{5}{2}, 0\right)$; vertex: (0, 25).**

You find the y-intercept by letting all the x's equal 0 and simplifying. Find x-intercepts by setting $25 - 4x^2$ equal to 0; the left side of the equation factors into $(5 - 2x)(5 + 2x) = 0$. You can then solve for x. The x-value of the vertex is $x = \dfrac{-0}{2(-4)} = 0$. You have no x term in the original formula (only x^2), so the coefficient b is equal to 0. Because the x-value of the coordinate for the vertex is 0, you get the y-intercept as the vertex.

3 Find the intercepts and vertex of the function $f(x) = x^2 + 10x + 7$. The answer is **y-intercept: (0, 7); x-intercepts: $\left(-5 + 3\sqrt{2}, 0\right), \left(-5 - 3\sqrt{2}, 0\right)$; vertex: (–5, –18).**

You find the y-intercept by letting all the x's equal 0 and simplifying. Find the x-intercepts by setting $x^2 + 10x + 7$ equal to 0 and solving with the quadratic formula (the trinomial doesn't factor):

$$x = \frac{-10 \pm \sqrt{10^2 - 4(1)(7)}}{2(1)} = \frac{-10 \pm \sqrt{100 - 28}}{2}$$

$$= \frac{-10 \pm \sqrt{72}}{2} = \frac{-10 \pm 6\sqrt{2}}{2} = -5 \pm 3\sqrt{2}$$

These x-intercepts end up being at about (–0.757, 0) and (–9.243, 0). The x-value of the vertex is $x = \dfrac{-(10)}{2(1)} = \dfrac{-10}{2} = -5$. Substituting the x-value in for all the x's, you get $f(-5) = (-5)^2 + 10(-5) + 7 = 25 - 50 + 7 = -18$. Thus, the vertex is at (–5, –18).

4 Find the intercepts and vertex of the function $f(x) = -x^2 - 8x - 16$. The answer is **y-intercept: (0, –16); x-intercept: (–4, 0); vertex: (–4, 0).**

You find the y-intercept by letting all the x's equal 0 and simplifying. Find the x-intercepts by setting $-x^2 - 8x - 16$ equal to 0, factoring that into $-1(x + 4)^2 = 0$, and solving for x. You find only one x-intercept; the graph of the parabola doesn't cross the axis at that intercept — it just touches (is tangent to) the axis. The x-value of the vertex is $x = \dfrac{-(-8)}{2(-1)} = \dfrac{8}{-2} = -4$. Substituting the x-value in for all the x's, you get $f(-4) = -(-4)^2 - 8(-4) - 16 = -16 + 32 - 16 = 0$. The vertex is at the x-intercept.

5 The height of a ball t seconds after it's thrown into the air from the top of a building can be modeled by $h(t) = -16t^2 + 48t + 64$, where $h(t)$ is height in feet. How high is the building, how high does the ball rise before starting to drop downward, and after how many seconds does the ball hit the ground? **The building is 64 feet tall, the ball peaks at 100 feet, and it takes 4 seconds to hit the ground.**

The ball is thrown from the top of the building, so you want the height of the ball when $t = 0$. This number is the initial t value (the y-intercept). When $t = 0$, $h = 64$, so the building is 64 feet high. The ball is at its highest at the vertex of the parabola. Calculating the t value, you get that the vertex occurs where $t = 1.5$ seconds. Substituting $t = 1.5$ into the formula, you get that $h = 100$ feet. The ball hits the ground when $h = 0$. Solving $-16t^2 + 48t + 64 = 0$, you factor to get $-16(t - 4)(t + 1) = 0$. The solution $t = 4$ tells you when the ball hits the ground. The $t = -1$ represents going backward in time, or in this case, where the ball would have started if it had been launched from the ground — not the top of a building.

6 The profit function telling Georgio how much money he will net for producing and selling x specialty umbrellas is given by $P(x) = -0.00405x^2 + 8.15x - 100$. What is Georgio's loss if he doesn't sell any of the umbrellas he produces, how many umbrellas does he have to sell to break even, and how many does he have to sell to earn the greatest possible profit? **Georgio loses $100 (earns –$100) if he sells 0, needs to sell 13 to break even, and can maximize profits if he sells 1,006 umbrellas.**

If Georgio sells no umbrellas, then $x = 0$, and he makes a negative profit (loss) of $100. The break-even point comes when the profit changes from negative to positive, at an x-intercept. Using the quadratic formula, you get two intercepts: at $x = 2,000$ and $x \approx 12.35$. The first (smaller) x-intercept is where the function changes from negative to positive. The second is where the profit becomes a loss again (too many umbrellas, too much overtime?). So, 13 umbrellas would yield a positive profit — he'd break even (have zero profit). The maximum profit occurs at the vertex. Using the formula for the x-value of the vertex, you get that $x \approx 1,006.17$. Substituting 1,006 into the formula, you get 4,000.1542; then substituting 1,007 into the formula, you get 4,000.15155. You see that Georgio gets slightly more profit with 1,006 umbrellas, but that fraction of a cent doesn't mean much. He'd still make about $4,000.

7 Chip ran through a maze in less than a minute the first time he tried. His times got better for a while with each new try, but then his times got worse (he took longer) due to fatigue. The amount of time Chip took to run through the maze on the ath try can be modeled by $T(a) = 0.5a^2 - 9a + 48.5$. How long did Chip take to run the maze the first time, and what was his best time? **Chip took 40 seconds the first time; his best time was 8 seconds.**

Because the variable a represents the number of the attempt, find $T(1)$ for the time of the first attempt. $T(1) = 40$ seconds. The best (minimum) time is at the vertex. Solving for the a value (which is the number of the attempt), $a = \dfrac{-(-9)}{2(0.5)} = 9$. He had the best time on the ninth attempt, and $T(9) = 8$.

8 A highway underpass is parabolic in shape. If the curve of the underpass can be modeled by $h(x) = 50 - 0.02x^2$, where x and $h(x)$ are in feet, then how high is the highest point of the underpass, and how wide is it? **The underpass is 50 feet high and 100 feet wide.**

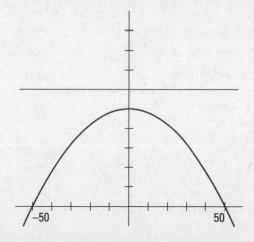

The highest point occurs at the vertex: $x = \dfrac{-0}{2(-0.02)} = 0$. The x-coordinate of the vertex is 0, so the vertex is also the y-intercept, at $(0, 50)$. The two x-intercepts represent the endpoints of the width of the overpass. Setting $50 - 0.02x^2$ equal to 0, you solve for x and get $x = 50, -50$. These two points are 100 units apart — the width of the underpass.

9 Sketch the graph of the parabola $f(x) = 4x^2$, labeling any intercepts and the vertex and showing the axis of symmetry.

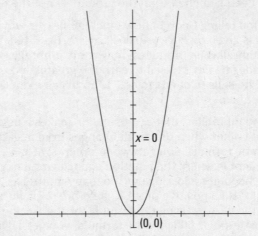

The only intercept is at $(0, 0)$. The parabola opens upward, because 4 is positive. The vertex is at $(0, 0)$, and the equation of the axis of symmetry is $x = 0$ (which is the y-axis).

10 Sketch the graph of the parabola $f(x) = -\dfrac{1}{3}x^2 + 3$, labeling any intercepts and the vertex and showing the axis of symmetry.

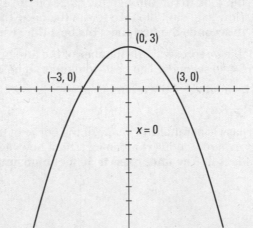

The intercepts are at $(0, 3)$, $(3, 0)$, and $(-3, 0)$. The parabola opens downward, because the coefficient of x^2 is negative. The vertex is at $(0, 3)$, the y-intercept, and the equation of the axis of symmetry is $x = 0$.

11 Sketch the graph of the parabola $f(x) = 3x^2 - 6x - 9$, labeling any intercepts and the vertex and showing the axis of symmetry.

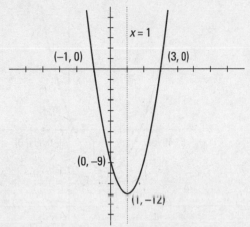

The intercepts are at $(0, -9)$, $(3, 0)$, and $(-1, 0)$. The parabola opens upward, because 3 is positive. The vertex is at $(1, -12)$, and the equation of the axis of symmetry is $x = 1$.

12 Sketch the graph of the parabola $f(x) = -2x^2 + 10x - 8$, labeling any intercepts and the vertex and showing the axis of symmetry.

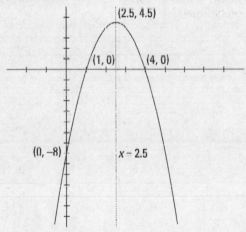

The intercepts are at $(0, -8)$, $(4, 0)$, and $(1, 0)$. The parabola opens downward because -2 is negative. The vertex is at $(2.5, 4.5)$, and the equation of the axis of symmetry is $x = 2.5$.

13 Sketch the graph of the parabola $f(x) = \frac{1}{2}x^2 + 3x + 1$, labeling any intercepts and the vertex and showing the axis of symmetry.

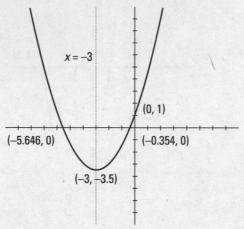

The y-intercept is at $(0, 1)$, and the x-intercepts are at approximately $(-0.354, 0)$ and $(-5.646, 0)$. The parabola opens upward, because $\frac{1}{2}$ is positive. The vertex is at $(-3, 3.5)$, and the equation of the axis of symmetry is $x = -3$.

14 Sketch the graph of the parabola $f(x) = -x^2 + 6x$, labeling any intercepts and the vertex and showing the axis of symmetry.

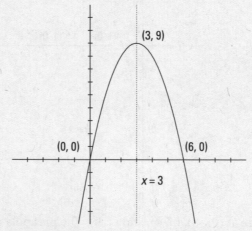

The intercepts are at $(0, 0)$ and $(6, 0)$; the intercept $(0, 0)$ is both the x-intercept and the y-intercept. The parabola opens downward because of the -1 coefficient on the x^2 term. The vertex is at $(3, 9)$, and the equation of the axis of symmetry is $x = 3$.

Chapter 7

Plugging In Polynomials

- -

In This Chapter

▶ Locating the intercepts in polynomials

▶ Figuring out where a polynomial function changes signs

▶ Assembling all the information you need to graph a polynomial curve

▶ Looking for roots with clues from Descartes — and others

▶ Doing division synthetically and taking other shortcuts

▶ Building up polynomials from their roots

- -

A polynomial function is a well-mannered function with nice, predictable habits. The graph of a polynomial function is always a smooth curve with no abrupt changes in direction or shape. The equation of a polynomial function has only whole-number exponents on the variables.

A *polynomial function* has the form $f(x) = a_n x^n + a_{n-1}x^{n-1} + a_{n-2}x^{n-2} + \cdots + a_1 x^1 + a_0 x^0$, where n is a whole number and any a_{n-i} is a real number. The x^0 is actually equal to 1, so that term is a constant; I include the final x and superscript here to complete the pattern. A polynomial function has at most nx-intercepts and at most $n - 1$ turning points (where the curve of the graph changes direction).

In this chapter, you graph polynomial functions after finding their intercepts and determining where the function is positive or negative. More advanced tools used to find the x-intercepts (roots) of the polynomial equation include the *rational root theorem* and *Descartes's rule of sign*. Finally, the *remainder theorem,* along with *synthetic division,* lets you quickly evaluate functions, and the factor theorem tells you how to reconstruct polynomials if you know only their roots.

Finding Basic Polynomial Intercepts

A polynomial function has exactly one y-intercept (where it crosses the y-axis). You find the y-intercept by replacing all the x's in the function formula with 0 and solving for y.

Polynomial functions may or may not have x-intercepts. If a polynomial function has an odd degree (the highest power is an odd number), then the polynomial has to have at least one x-intercept; the curve has to cross the x-axis. A polynomial function with an even degree may or may not cross the x-axis. But in any case, a polynomial won't have more x-intercepts than its degree. You can bank on that.

To solve for the x-intercepts, set the function equation equal to 0 and find out which values of x are solutions. Those answers are the x-values of the intercepts. The biggest challenge often comes when trying to find that solution of the function equation. The problems in this first section are easily solvable. The next section and "Possible Roots and Where to Find Them: The Rational Root Theorem and Descartes's Rule" show you how to handle equations that aren't as cooperative with their solutions.

Q. Find the intercepts of the polynomial $y = x^3 - 16x$.

A. **y-intercept: (0, 0); x-intercepts: (0, 0), (4, 0), and (–4, 0).** Note the repeat in one answer. The intercept (0, 0) always appears in both intercept categories (or in neither). To find the y-intercept, replace all the x's in the function equation with 0 and solve: $y = 0^3 - 16(0) = 0$. To find the x-intercepts, set the function equation equal to 0 and solve: $0 = x^3 - 16x = x(x^2 - 16) = x(x - 4)(x + 4)$. The solutions are $x = 0$, 4, and –4.

Q. Find the intercepts of the polynomial $y = x^5 + 3x^4 - 5x^3 - 15x^2$.

A. **y-intercept: (0, 0); x-intercepts: (0, 0), (–3, 0), $\left(\sqrt{5}, 0\right), \left(-\sqrt{5}, 0\right)$.** You might expect to find as many as five x-intercepts for this function, but as it turns out, you have only four. The rule gives you the maximum number of x-intercepts possible, not always exactly how many. Replacing all the x's with 0's and simplifying tells you that the y-intercept is 0. To solve for the x-intercepts, you have to factor the function rule. First, pull out the x^2 factor, and then factor the four terms in the parentheses by grouping (Chapter 2 explains this factoring technique): $x^5 + 3x^4 - 5x^3 - 15x^2 = x^2(x^3 + 3x^2 - 5x - 15) = x^2[x^2(x + 3) - 5(x + 3)] = x^2[(x + 3)(x^2 - 5)]$. The exponent 2 on the x term indicates a double root at $x = 0$ (I explain how a double root affects the graph in "Graphing Polynomials," later in this chapter). The factor $x^2 - 5$ yields two roots — one positive and one negative — because you have to take a square root to solve for x.

1. Find the intercepts of the polynomial $y = (x + 4)(x + 2)(x^2 - 1)$.

2. Find the intercepts of the polynomial $y = x(x - 3)^2(x - 6)(x^2 + 4)$.

3. Find the intercepts of the polynomial $y = -2x^3 - 6x^2 + 8x$.

4. Find the intercepts of the polynomial $y = x^4 - 26x^2 + 25$.

Digging Up More-Difficult Polynomial Roots with Factoring

When solving for roots (*x*-intercepts of a polynomial), you usually need to factor the function rule and set it equal to 0. The factorization can be simple and obvious or complicated and obscure. You always hope for the simple and obvious, move to challenging and doable, and resort to the "big guns" when the factors are more obscure. You can find the "big guns" later in this chapter in "Possible Roots and Where to Find Them: The Rational Root Theorem and Descartes's Rule."

Before you go to those lengths, though, you need to exhaust other methods. The other methods of factoring include

- ✔ Dividing out a greatest common factor (GCF)

- ✔ Factoring a perfect square binomial

- ✔ Factoring by grouping

- ✔ Factoring quadratic-like trinomials

Flip to Chapter 2 if you need a review of these methods.

Q. Find the roots (solutions) of the polynomial $x^7 - 82x^5 + 81x^3 = 0$.

A. $x = 0, 0, 0, 9, -9, 1, -1$. This equation technically has seven solutions, but the 0 is a multiple root, so you end up with only five different numbers. To find these solutions, you first factor x^3 out of each term to get $x^3 \left(x^4 - 82x^2 + 81 \right) =$. The two binomials are both the difference of perfect squares, so you can factor them into the difference and sum of the roots of the terms. You get $x^3(x - 9)(x + 9)(x - 1)(x + 1) = 0$. Setting each of the factors equal to 0, you find the roots.

Q. Find the roots (solutions) of the polynomial $x^3 - 16x^2 + 100x - 1,600 = 0$.

A. $x = 16$. The polynomial doesn't have a common factor in the four terms, but you can group the terms for the pairs of common factors. You get $x^2(x - 16) + 100(x - 16) = 0$, which factors into $(x - 16)(x^2 + 100) = 0$. The second binomial is the sum of squares, which doesn't factor. Setting these two factors equal to 0, you get $x = 16$ from the first factor, but the second factor doesn't produce any real answers. Even though you started with a third-degree polynomial, which can yield up to three solutions, this polynomial has only one real root. The solution is just $x = 16$.

5. Find the roots (solutions) of the polynomial $3x^4 - 12x^3 - 27x^2 + 108x = 0$.

6. Find the roots (solutions) of the polynomial $x^5 - 16x^3 + x^2 - 16 = 0$.

7. Find the roots (solutions) of the polynomial $x^6 + 9x^3 + 8 = 0$.

8. Find the roots (solutions) of the polynomial $36x^5 - 13x^3 + x = 0$.

Determining Where a Function Is Positive or Negative

A polynomial function is positive when the function value (or y-value) is a positive number. Graphically, you can see where a polynomial function is positive by where the curve is above the x-axis. When you don't have a graph to look at, one way to determine where a polynomial function is positive is to just plug numbers in for x and see what you get. However, not only is plugging in random numbers time-consuming, but it also can lead to inaccuracies.

The best method for determining on what intervals (between what x-values) a polynomial is positive or negative is to find the *zeros* or *roots* (x-intercepts; values for which the polynomial is 0), and then plug in a *test number* between each pair of intercepts; this test number lets you check the positive or negative nature of the function there. Make sure you also use test numbers in the intervals from $-\infty$ to the lowest x-intercept and from the highest x-intercept to ∞.

To write the positive and negative natures of a function in "mathspeak," you can use inequalities or interval notation. Remember that at x-intercepts, the function equals 0. Because 0 is neither positive nor negative, you can't use $\leq$, $\geq$, or brackets when you write your answers; answers simply include $<$ and $>$ or parentheses when next to infinity. (For details on interval notation, see Chapter 1.)

Q. Determine where the polynomial $y = x(x - 3)(x - 7)(x + 2)^2(x + 6)$ is positive and where it's negative.

A. **Positive: $(-\infty, -6)$, $(0, 3)$, $(7, \infty)$; negative: $(-6, -2)$, $(-2, 0)$, $(3, 7)$.** You first determine the x-intercepts (zeros or roots) by setting the function equal to 0. The polynomial $x(x - 3)(x - 7)(x + 2)^2(x + 6) = 0$ when $x = 0$, $x = 3$, $x = 7$, $x = -2$, and $x = -6$. Draw a number line with the x-values representing the zeros shown on the line — in their correct order from smallest to largest. Figure 7-1 shows the bare number line on the top and the "worked through" number line underneath that first one.

You test each interval sectioned off by the zeros on the number line by choosing some *test number* that lies in each interval and plugging that number in for each x in the function rule. You determine whether each individual factor is positive or negative, record that fact on the number line, and then see whether the product of all those factors comes out to be positive or negative. For instance, if you look at the interval between the -2 and 0 on the number line, the most obvious choice for x is -1. (Any number works — pick a nice one when you can.) Replacing all the x's in

the function rule with -1, you see that the first factor, x, is negative, the factor $(x - 3)$ is negative, the $(x - 7)$ is negative, the $(x + 2)^2$ is positive, and the $(x + 6)$ is positive. You write $(-)(-)(-)(+)(+)$ in the interval and determine that the product is negative (if you have an odd number of negative signs, the product is negative; an even number of negative signs, and the product is positive). So the function is negative for all the numbers between -2 and 0.

Reading the signs (sort of like astrology?) on the number line, you determine that the value of the function is positive when the x's are less than -6, between 0 and 3, or greater than 7. Between -6 and -2, between -2 and 0, or between 3 and 7, the function is negative. Even though you don't have a sign change at -2, you can't include -2 in the numbers that give a negative function value. At -2, the function is 0. Here's the answer in inequality notation: the function is positive for $x < -6$ or $0 < x < 3$ or $x > 7$, and the function is negative for $-6 < x < -2$ or $-2 < x < 0$ or $3 < x < 7$. In interval notation, you write that the function is positive for $(-\infty, -6) \cup (0, 3) \cup (7, \infty)$, and it's negative for $(-6, -2) \cup (-2, 0) \cup (3, 7)$. The $\cup$ symbol means *union* and is standard set notation. It replaces the word *or*.

Figure 7-1:
A number line shows the positive and negative intervals.

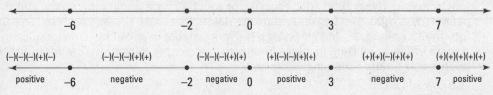

9. Determine where the polynomial $y = 6x^2 - 6x$ is positive and where it's negative.

10. Determine where the polynomial $y = -2x^2 - 7x + 15$ is positive and where it's negative.

11. Determine where the polynomial $y = (x + 3)(x - 2)(x - 5)^2$ is positive and where it's negative.

12. Determine where the polynomial $y = x^2(x + 3)^2(x - 4)^4$ is positive and where it's negative.

Graphing Polynomials

The graph of a polynomial is a smooth, winding curve that starts somewhere to the left, moves across the graph paper, crossing or touching or avoiding the x-axis, and then disappears to the right. You can determine the points that make up the graph by inserting x-values into the function rule, but graphing a polynomial that way can be a long, tedious chore. A more efficient method involves finding the intercepts, plotting a few additional points for clarity, and boldly sketching in the rest.

Q. Sketch the graph of $y = (x^2 - 1)(x + 3)^2$.

A. First, determine the intercepts and place them on the graph. When $x = 0$ in the function equation, $y = -9$; that's the y-intercept. When $y = 0$, $x = 1, -1,$ or -3; those are the three x-intercepts. The intercept at $(-3, 0)$ is a double root. The curve only touches the axis at that point — it doesn't cross the x-axis. Next, use a few test points between the intercepts to figure out where the function is positive or negative. The function is negative between $x = -1$ and $x = 1$. Other than that (and at the x-intercepts), the function is positive — above the x-axis. Figure 7-2 shows you the completed graph. The graph shows the intercepts and a few arbitrary points.

Q. Sketch the graph of $y = -x^3 - x^2 + 25x + 25$.

A. The y-intercept is at (0, 25). You find that answer by letting all the x's be 0. To solve for the x-intercepts, set the function rule equal to 0 and factor by grouping: $0 = -x^2(x + 1) + 25(x + 1) = (x + 1)(25 - x^2) =$ $(x + 1)(5 - x)(5 + x)$. The x-intercepts are (–1, 0), (5, 0), and (–5, 0). The function is negative on the intervals (–5, –1) and (5, ∞) and positive on the intervals (–∞, –5) and (–1, 5). Figure 7-3 shows the graph of the function with the intercepts and a few select points for clarity.

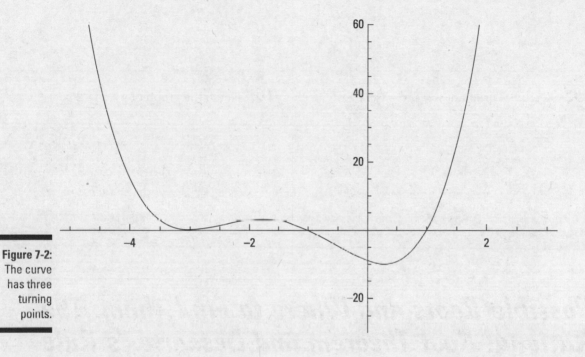

Figure 7-2:
The curve has three turning points.

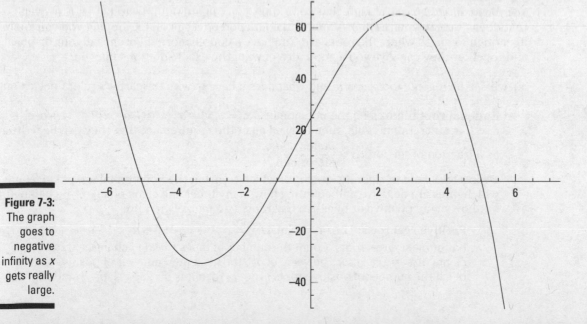

Figure 7-3:
The graph goes to negative infinity as x gets really large.

13. Sketch the graph of $y = (x + 4)(x + 2)(x - 3)$.

14. Sketch the graph of $y = x(x - 3)^2(x - 6)$.

15. Sketch the graph of $y = -2x^3 - 6x^2 + 8x$.

16. Sketch the graph of $y = x^4 - 26x^2 + 25$.

Possible Roots and Where to Find Them: The Rational Root Theorem and Descartes's Rule

You may come across problems that you simply can't figure out how to factor. However, that doesn't mean you're stuck graphing the function on a calculator, hoping you can zoom in enough to guess where the roots are. You have a couple more theorems at your disposal, and together they can give you a good idea of what the exact solutions may be.

So, what's the *rational root theorem* and what does it have to with Descartes's rule? Check it out:

✔ **Rational root theorem:** If the polynomial $f(x) = a_nx^n + a_{n-1}x^{n-1} + a_{n-2}x^{n-2} + \cdots + a_1x^1 + a_0x^0$ has any rational roots, then they all meet the requirement that they can be written as a fraction of the form $x = \dfrac{\text{factor of } a_0}{\text{factor of } a_n}$.

✔ **Descartes's rule of sign:** By looking at which terms in the polynomial are positive and which ones are negative and counting how many times the signs change, you can figure out how many positive or negative real roots you're likely to find:

  • **Positive real roots:** The polynomial $f(x) = a_nx^n + a_{n-1}x^{n-1} + a_{n-2}x^{n-2} + \cdots + a_1x^1 + a_0x^0$ has at most n real roots. Count the number of times the sign changes in the coefficients of f, and call it p. The value of p is the maximum number of positive real roots of f. If the number of positive roots is not p, then it is $p - 2, p - 4, p - 6$, and so on.

- **Negative real roots:** In the polynomial $f(x) = a_n x^n + a_{n-1}x^{n-1} + a_{n-2}x^{n-2} + \cdots + a_1 x^1 + a_0 x^0$ find $f(-x)$. Then count the number of times the sign changes in $f(-x)$ and call it q. The value of q is the maximum number of *negative* real roots of f. If the number of negative roots is not q, then it is $q - 2$, $q - 4$, $q - 6$, and so on.

If the polynomial doesn't have a constant term, you first have to factor out the greatest power of x possible before applying the rational root theorem or Descartes's rule of sign.

Q. Determine all the possibilities for rational roots of the polynomial $4x^4 - 25x^2 + 36 = 0$. Then determine how many of the real roots of the polynomial may be positive and how many may be negative. Factor the polynomial to confirm your results.

A. **Possible rational roots:** $\pm 1, \pm 2, \pm 3, \pm 4, \pm 6, \pm 9, \pm 12, \pm 18, \pm 36, \pm \frac{1}{2}, \pm \frac{3}{2}, \pm \frac{9}{2}, \pm \frac{1}{4}, \pm \frac{3}{4}, \pm \frac{9}{4}$;

number of positive real roots: two or zero; number of negative real roots: two or zero;

actual solution: $x = \frac{3}{2}, -\frac{3}{2}, 2, -2$. Any rational roots of this polynomial equation have to be numbers that are the results of dividing factors of the constant term, 36, by factors of the lead coefficient, 4. First, list the factors of the constant term. Use both + and – for each factor. The list reads: $\pm 1, \pm 2, \pm 3, \pm 4, \pm 6, \pm 9, \pm 12, \pm 18, \pm 36$. These numbers are all rational, and they're all possible solutions of the equation. Now divide each of the factors of 36 by the factors of 4. The factors of 4 are $\pm 1, \pm 2, \pm 4$. Dividing the factors of 36 by ± 1 won't change any of them, so don't bother. Dividing the factors by ± 2 and ± 4 gives you $\frac{\pm 1}{\pm 2}, \frac{\pm 2}{\pm 2}, \frac{\pm 3}{\pm 2}, \frac{\pm 4}{\pm 2}, \frac{\pm 6}{\pm 2}, \frac{\pm 9}{\pm 2}, \frac{\pm 12}{\pm 2}, \frac{\pm 18}{\pm 2}, \frac{\pm 36}{\pm 2}$ and $\frac{\pm 1}{\pm 4}, \frac{\pm 2}{\pm 4}, \frac{\pm 3}{\pm 4}, \frac{\pm 4}{\pm 4}, \frac{\pm 6}{\pm 4}, \frac{\pm 9}{\pm 4}, \frac{\pm 12}{\pm 4}, \frac{\pm 18}{\pm 4}, \frac{\pm 36}{\pm 4}$.

Reducing these fractions and eliminating the duplicates, the remaining fractions are $\frac{\pm 1}{\pm 2}, \frac{\pm 3}{\pm 2}, \frac{\pm 9}{\pm 2}, \frac{\pm 1}{\pm 4}, \frac{\pm 3}{\pm 4}, \frac{\pm 9}{\pm 4}$. The list of integers plus these fractions are all the possibilities for rational roots. No other rational numbers can work in this function.

Now for the positive and negative natures of the real roots: First look at $f(x) = 4x^4 - 25x^2 + 36$ and count the number of sign changes from beginning to end. The terms change from positive (+4) to negative (–25) back to positive (+36). You find two sign changes, so the equation has either two or zero positive real roots. Notice that the two exponents are both even (and technically, the 36 is multiplied by x raised to the 0 power — which is also an even exponent). Now look at $f(-x) = 4(-x)^4 - 25(-x)^2 + 36 = 4x^4 - 25x^2 + 36$. The signs didn't change; you still have two sign changes, so there are either two or zero negative real roots. Now, to check the answer, the factorization is $4x^4 - 25x^2 + 36 = (4x^2 - 9)(x^2 - 4) = (2x - 3)(2x + 3)(x - 2)(x + 2) = 0$. The solutions are $x = \frac{3}{2}, -\frac{3}{2}, 2, -2$, so you find exactly two positive and two negative real roots.

17. Determine all the possibilities for rational roots of the polynomial $9x^4 - 37x^2 + 4 = 0$. Then determine how many of the real roots of the polynomial may be positive and how many may be negative. Factor the polynomial to confirm your results.

18. Determine all the possibilities for rational roots of the polynomial $2x^4 - 3x^3 - 18x^2 + 27x = 0$. Then determine how many of the real roots of the polynomial may be positive and how many may be negative. Factor the polynomial to confirm your results.

19. Determine all the possibilities for rational roots of the polynomial $x^4 - 4x^3 + 6x^2 - 4x + 1 = 0$. Then determine how many of the real roots of the polynomial may be positive and how many may be negative. Factor the polynomial to confirm your results.

20. Determine all the possibilities for rational roots of the polynomial $x^5 + 8x^4 + 7x^3 - 25x^3 - 200x^2 - 175x = 0$. Then determine how many of the real roots of the polynomial may be positive and how many may be negative. Factor the polynomial to confirm your results.

21. Determine all the possibilities for rational roots of the polynomial $3x^3 + 4x^2 + 30x + 40 = 0$. Then determine how many of the real roots of the polynomial may be positive and how many may be negative. Factor the polynomial to confirm your results.

22. Determine all the possibilities for rational roots of the polynomial $x^6 - 16x^3 + 64 = 0$. Then determine how many of the real roots of the polynomial may be positive and how many may be negative. Factor the polynomial to confirm your results.

Getting Real Results with Synthetic Division and the Remainder Theorem

Synthetic division is a method for dividing polynomials by binomials. Because this type of division occurs frequently in algebra and other mathematics courses, the process is especially useful. Synthetic division, along with the remainder theorem, can help you quickly evaluate a polynomial function at any value.

Basically, synthetic division lets you cut out all the "middlemen." You deal with the coefficients of the terms in the polynomial and the constant in the binomial and ignore (well, ignore just momentarily) the variables while doing the process. The easiest way to explain synthetic division is just to show you an example.

The remainder theorem says that when the polynomial $f(x) = a_n x^n + a_{n-1} x^{n-1} + a_{n-2} x^{n-2} + \cdots + a_1 x^1 + a_0$ is divided by the binomial $x - c$, the remainder of the division is equal to $f(c)$.

For instance, when you divide $f(x) = x^3 + 2x^2 - 5x + 4$ by the binomial $x - 1$, the remainder is 2, and $f(1) = 2$. You can find $f(1)$ by using 1 as your divisor in synthetic division.

Q. Do the division problem $(4x^5 + 3x^4 - 6x^2 + 5x - 2) \div (x + 3)$ using synthetic division.

A. $4x^4 - 9x^3 + 27x^2 + 266 - \dfrac{800}{x+3}$. Here's how synthetic division works:

1. Set up the problem.

   Put the terms of the polynomial in decreasing order of the exponents, inserting a 0 for any missing term. Then write the coefficients of the variables and the constant in a horizontal row. Put a division symbol, a little half-box, to the left of that row. The number you put in front of the division symbol (the divisor) is the *opposite* of the number in the binomial you're dividing by. So, if you divide $4x^5 + 3x^4 - 6x^2 + 5x - 2$ by the binomial $x + 3$, you use -3 in synthetic division. Here's the initial setup of the division problem:

   $$-3\underline{\big|\;\; 4 \quad 3 \quad 0 \quad -6 \quad 5 \quad -2}$$

   Notice that you write a 0 where the coefficient of the x^3 term would be. The horizontal line leaves enough room beneath the numbers to write products.

2. Drop the lead coefficient below the horizontal line.

   In this case, the lead coefficient is the 4.

3. Multiply the number below the horizontal line by the divisor; write this product beneath the next coefficient, and add the coefficient and product together; put your answer below the horizontal line, and repeat the process.

   Multiply the 4 by -3 and put that product, -12, beneath the next coefficient, the 3. Add the 3 and -12, and put that result, -9, below the horizontal line. Multiply -9 by -3, put the product under the next coefficient, the 0, and add the 0 and 27. Keep repeating the multiply-add process all the way down the line until you're done:

   $$
   \begin{array}{r|rrrrrr}
   -3 & 4 & 3 & 0 & -6 & 5 & 2 \\
      &   & -12 & 27 & -81 & 261 & -798 \\
   \hline
      & 4 & -9 & 27 & -87 & 266 & -800
   \end{array}
   $$

4. Put the variables back in as you write the answer — lowering the exponent of the first variable by 1.

   The last entry on the bottom is -800. This number is the remainder of the division, and the other numbers are coefficients of the quotient (answer). To write out the answer, start with the fourth power (one lower than the original fifth power), and use all the coefficients, dropping the power on the variable by 1 with each successive coefficient. The last number is the remainder, so write that over the divisor. In this case, write the answer of the division problem as $4x^4 - 9x^3 + 27x^2 + 266 + \dfrac{-800}{x+3}$.

Q. Using the remainder theorem, find $f(-9)$ if $f(x) = x^5 + 10x^4 + 7x^3 - 20x^2 - 18x + 4$.

A. $f(-9) = 4$. If you substitute the -9 in for the x's and simplify, you get 4 — if you're lucky. The numbers get really large and unwieldy. Using the remainder theorem and synthetic division, you have a much easier task: Use -9 as your divisor when you do synthetic division; your answer is the remainder, the last entry on the bottom:

$$
\begin{array}{r|rrrrrr}
-9 & 1 & 10 & 7 & -20 & -18 & 4 \\
   &   & -9 & -9 & 18 & 18 & 0 \\
\hline
   & 1 & 1 & -2 & -2 & 0 & 4
\end{array}
$$

23. Use synthetic division to divide $(9x^4 - 37x^2 + 4) \div (x - 2)$.

24. Use synthetic division to divide $(2x^4 + 3x^3 + x^2 - 3x + 5) \div (x + 1)$.

25. Use synthetic division to divide $(3x^3 + 4x^2 - 2x + 10) \div (x + 2)$.

26. Use synthetic division to divide $(x^4 - 12x^3 + 54x^2 - 108x + 81) \div (x - 3)$.

27. Find $f(2)$ if $f(x) = x^5 - 3x^4 + 4x^3 - 6x^2 + 5x - 1$.

28. Find $f(-2)$ if $f(x) = 6x^{10} - 16x^6 + 150x^5 - 80x^2$.

Connecting the Factor Theorem with a Polynomial's Roots

Factors of polynomials and roots of polynomial equations are closely linked. Whenever a variable is involved, factors produce roots, and roots identify factors.

The factor theorem is special because it works forward and backward. Technically, the theorem is an *if and only if* type of statement. *If* inputting some number c makes the function equal to 0, *then* it's true that $x - c$ is a factor of the function. And *if* you know that $x - c$ is a factor of some function, *then* inputting c must make the function equal 0.

Factor theorem says that if $x = \frac{b}{a}$ is a root of the polynomial $f(x)$, then the binomial $(ax - b)$ is a factor of the polynomial function $f(x)$. Also, if a polynomial factors into $f(x) = k(x - a)$ $(x - b) \ldots$ where k is some constant factor, then a is a root, b is a root, and so on.

For instance, the polynomial $f(x) = 2x^3 - x^2 - 18x + 9$ has the roots $x = \frac{1}{2}$, 3, –3, and it factors into $f(x) = (2x - 1)(x - 3)(x + 3)$.

Q. If the roots of $f(x)$ are $x = 2, -4, 6, -8, -8, \frac{2}{3}$, write a function rule for $f(x)$.

A. **Using the factor theorem, $f(x) = k(x - 2)$ $(x + 4)(x - 6)(x + 8)^2(3x - 2)$.** You're probably wondering what that k is for. With the information given, you can write only the factors that involve variables. Each factor, when set equal to 0, gives you one of the roots (solutions) of the polynomial equation. But the original function could contain some constant multiplier that doesn't show up in the roots. The value of the constant doesn't really matter unless you're graphing the function. To figure out what k is, you need more information. For instance, if you know that $f(0) = 6,144$, you can determine that $k = 1$. For now, just put the k in as a placeholder and as a reminder that something may be missing from the big picture.

---

29. If the roots of $f(x)$ are $x = 1, 2, -2, 3, 3, 3, 4, 4$, then write a function rule for $f(x)$.

30. If the roots of $f(x)$ are $x = 0, 0, 0, 1, 1, 1, 1, -2, -3, -4$, then write a function rule for $f(x)$.

---

31. If the roots of $f(x)$ are $x = -1, -1, 1, \frac{1}{2}, \frac{3}{4}, -\frac{7}{8}, -\frac{7}{8}$, then write a function rule for $f(x)$.

Answers to Problems on Plugging In Polynomials

The following are the answers to the practice problems presented earlier in this chapter.

1 Find the intercepts of the polynomial $y = (x + 4)(x + 2)(x^2 - 1)$. The answer is **y-intercept: (0, –8); x-intercepts: (–4, 0), (–2, 0), (1, 0), (–1, 0).**

Replace all the x's with 0 to get the y-intercept, $(0, –8)$. You can obtain the x-intercepts by letting $y = 0$ and setting each factor equal to 0 to get the solutions.

2 Find the intercepts of the polynomial $y = x(x - 3)^2(x - 6)(x^2 + 4)$. The answer is **y-intercept: (0, 0); x-intercepts: (0, 0), (3, 0), (6, 0).**

Replace all the x's with 0 to get the y-intercept, $(0, 0)$. This answer is also, technically, an x-intercept. Obtain the x-intercepts by letting $y = 0$ and setting each factor equal to 0. The factor $(x^2 + 4)$ never equals 0, so you don't get an x-intercept from that factor.

3 Find the intercepts of the polynomial $y = -2x^3 - 6x^2 + 8x$. The answer is **y-intercept: (0, 0); x-intercepts: (0, 0), (–4, 0), (1, 0).**

First, factor the polynomial to get $y = -2x(x + 4)(x - 1)$. Replacing the x's with 0 gives you the y-intercept $(0, 0)$, which is also an x-intercept. Setting the binomials equal to 0 gives you the other two x-intercepts.

4 Find the intercepts of the polynomial $y = x^4 - 26x^2 + 25$. The answer is **y-intercept: (0, 25); x-intercepts: (5, 0), (–5, 0), (1, 0), (–1, 0).**

Factoring the polynomial, you get $y = (x^2 - 25)(x^2 - 1) = (x - 5)(x + 5)(x - 1)(x + 1)$. Replacing all the x's with zeros gives you the y-intercept $(0, 25)$. Setting each binomial factor equal to 0 gives you all the x-intercepts.

5 Find the roots (solutions) of the polynomial $3x^4 - 12x^3 - 27x^2 + 108x = 0$. The answer is **x = 0, 3, –3, 4.**

First factor $3x$ out of each term to get $3x(x^3 - 4x^2 - 9x + 36) = 0$. You can then factor the terms in the parentheses by grouping: $3x[x^2(x - 4) - 9(x - 4)] = 3x[(x - 4)(x^2 - 9)] = 3x[(x - 4)(x - 3)(x + 3)] = 0$. Set each factor equal to 0 to solve for the roots.

6 Find the roots (solutions) of the polynomial $x^5 - 16x^3 + x^2 - 16 = 0$. The answer is **x = 4, –4, –1.**

The polynomial factors by grouping: $x^3(x^2 - 16) + 1(x^2 - 16) = (x^2 - 16)(x^3 + 1) = (x - 4)(x + 4)(x + 1)(x^2 - x + 1) = 0$. The first three factors give you the real roots. The last factor is a quadratic that has no real solution when you set it equal to 0.

Note: You don't really have to factor $x^3 + 1$ in problem 6 to find the root. If you just set $x^3 + 1$ equal to 0, you get $x^3 = -1$, and taking the cube root of both sides gives you the solution –1. The factored form simply shows you how this problem could have had five roots, but not all are real numbers in this case.

7 Find the roots (solutions) of the polynomial $x^6 + 9x^3 + 8 = 0$. The answer is **x = –2, –1.**

The polynomial is quadratic-like (see Chapter 2 if you need more information on factoring quadratic-like trinomials). It factors into $(x^3 + 8)(x^3 + 1) = 0$. Setting each factor equal to 0, you get the two roots.

8 Find the roots (solutions) of the polynomial $36x^5 - 13x^3 + x = 0$. The answer is $x = 0, \frac{1}{3}, -\frac{1}{3}, 2, -2$.

First, factor x out of each term to get $x(36x^4 - 13x^2 + 1) = 0$. The quadratic-like trinomial factors, giving you $x(9x^2 - 1)(4x^2 - 1) = 0$. Each binomial is the difference of squares (see Chapter 2), so both binomials factor. For the final factorization, you end up with $x(3x - 1)(3x + 1)(2x - 1)(2x + 1) = 0$. Setting each factor equal to 0 gives you the five different solutions.

9 Determine where the polynomial $y = 6x^2 - 6x$ is positive and where it's negative. The answer is **positive: $(-\infty, 0), (1, \infty)$; negative: $(0, 1)$.**

First, factor the polynomial to get $6x(x - 1)$. The zeros are at 0 and 1. Set up a number line with the 0 and 1, placed in that order. Check the signs of the factors in the three different intervals: less than 0, between 0 and 1, and then greater than 1. The products of the factors are negative for all values of x between 0 and 1 and positive everywhere else.

10 Determine where the polynomial $y = -2x^2 - 7x + 15$ is positive and where it's negative. The answer is **positive: $\left(-5, \frac{3}{2}\right)$; negative: $(-\infty, -5) \cup \left(\frac{3}{2}, \infty\right)$.**

Factor the polynomial to get $y = -(2x - 3)(x + 5)$. Create a number line with the roots in their correct order. Checking the signs in the three intervals, you find that the function is positive only between the two zeros and negative elsewhere.

11 Determine where the polynomial $y = (x + 3)(x - 2)(x - 5)^2$ is positive and where it's negative. The answer is **positive: $(-\infty, -3)$, $(2, 5)$, $(5, \infty)$; negative: $(-3, 2)$.**

Place the numbers representing the roots or zeros on the number line; then check out the signs of the factors and products of the factors to determine the sign of the product in each interval.

12 Determine where the polynomial $y = x^2(x + 3)^2(x - 4)^4$ is positive and where it's negative. The answer is **positive: $(-\infty, -3), (-3, 0), (0, 4), (4, \infty)$; negative: never.**

It's probably easiest just to say that this polynomial is positive *except* at the three zeros. The even powers make the binomials positive, except when they're 0.

13 Sketch the graph of $y = (x + 4)(x + 2)(x - 3)$.

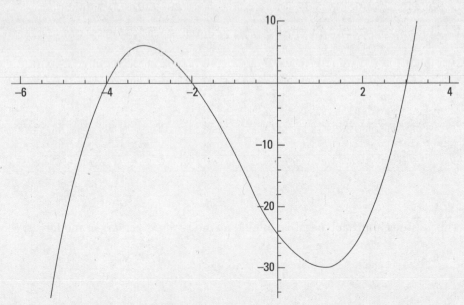

The graph continues on to get higher and higher as x gets larger and larger.

14 Sketch the graph of $y = x(x-3)^2(x-6)$.

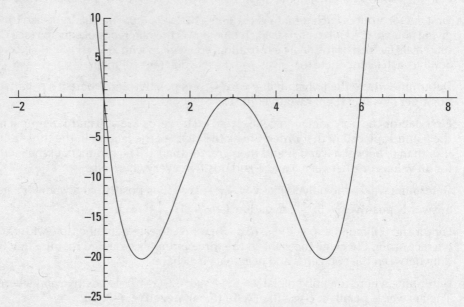

A double root is at $x = 3$, which is why the curve doesn't cross the axis at that point.

15 Sketch the graph of $y = -2x^3 - 6x^2 + 8x$.

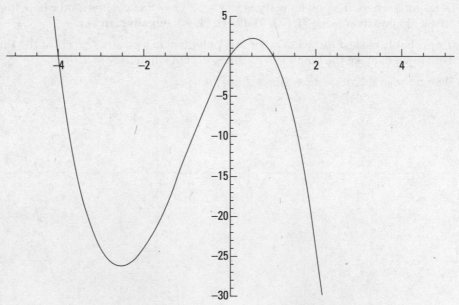

The y-values approach negative infinity as the x-values get larger and larger.

16 Sketch the graph of $y = x^4 - 26x^2 + 25$.

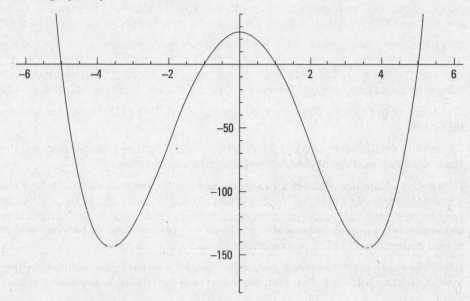

The curve is symmetric with respect to the y-axis because it's an even function. (Chapter 5 can tell you more about even and odd functions.)

17 Determine all the possibilities for rational roots of the polynomial $9x^4 - 37x^2 + 4 = 0$. Then determine how many of the real roots of the polynomial may be positive and how many may be negative. Factor the polynomial to confirm your results. The answer is **possible rational roots:** $\pm 1, \pm 2, \pm 4, \pm\frac{1}{3}, \pm\frac{2}{3}, \pm\frac{4}{3}, \pm\frac{1}{9}, \pm\frac{2}{9}, \pm\frac{4}{9}$; **number of possible real roots — positive: two or zero, negative: two or zero: actual roots:** $x = \frac{1}{3}, -\frac{1}{3}, 2, -2$.

Use the rational root theorem to determine all the possible rational roots. Put the factors of the constant, ± 1, ± 2, and ± 4, over the factors of the lead coefficient, 1, 3, and 9. You don't really have to divide by the 1, because that doesn't change the value.

The signs change twice in the function rule, so you can have two or zero positive real roots. When you replace all the x's with $-x$, the function rule doesn't change at all (because of the even exponents). Again, you find two or zero negative real roots. Factoring the polynomial, you get $(3x - 1)(3x + 1)(x - 4)(x + 4) = 0$. Setting the factors equal to 0, you get the roots $x = \frac{1}{3}, -\frac{1}{3}, 2, -2$. This function has exactly two positive and two negative real roots. Notice that all four roots are in the list created using the rational root theorem.

18 Determine all the possibilities for rational roots of the polynomial $2x^4 - 3x^3 - 18x^2 + 27x = 0$. Then determine how many of the real roots of the polynomial may be positive and how many may be negative. Factor the polynomial to confirm your results. The answer is **possible rational roots:** ± 1, ± 3, ± 9, ± 27, $\pm\frac{1}{2}$, $\pm\frac{3}{2}$, $\pm\frac{9}{2}$, $\pm\frac{27}{2}$; **number of possible real roots — positive: two or zero, negative: one; actual roots:** $0, \frac{3}{2}, 3, -3$.

First, factor out the x from each term to get $x(2x^3 - 3x^2 - 18x + 27) = 0$. Then use the rational root theorem on the terms in the parentheses to determine all the possible rational roots. Put the factors of the constant, ±1, ±3, ±9, and ±27, over the factors of the lead coefficient, 1 and 2.

You have two sign changes on the terms within the parentheses. Again, look at the polynomial inside the parentheses, and change each x to a $-x$. You get $2(-x)^3 - 3(-x)^2 - 18(-x) + 27 = -2x^3 - 3x^2 + 18x + 27$. You find only one sign change. Factoring the polynomial inside the parentheses by grouping, $x[x^2(2x - 3) - 9(2x - 3)] = x[(2x - 3)(x^2 - 9)] = x[(2x - 3)(x - 3)(x + 3)] = 0$.

The roots of the entire polynomial are $x = 0, \frac{3}{2}, 3, -3$. Two of the roots are positive; one is negative.

The number 0 is also a root, but 0 has no sign — it's neither positive nor negative, so it doesn't show up when you count possible positive and negative roots.

19 Determine all the possibilities for rational roots of the polynomial $x^4 - 4x^3 + 6x^2 - 4x + 1 = 0$. Then determine how many of the real roots of the polynomial may be positive and how many may be negative. Factor the polynomial to confirm your results. The answer is **possible rational roots: ±1; number of possible real roots — positive: four or two or zero, negative: zero; actual roots: $x = 1, 1, 1, 1$ (a quadruple root).**

Using the rational root theorem, you divide the factors of the constant, 1, by the factors of the lead coefficient, also a 1. That step gives you only two different possibilities for rational roots: 1 and -1.

The signs change four times in the original polynomial, indicating 4 or 2 or 0 positive real roots. Replacing each x with $-x$, you get $x^4 + 4x^3 + 6x^2 + 4x + 1 = 0$. The signs never change. The polynomial is the fourth power of the binomial $(x - 1)$, so it factors into $(x - 1)^4 = 0$, and the roots are 1, 1, 1, and 1. There are four positive roots (all the same number, of course).

20 Determine all the possibilities for rational roots of the polynomial $x^5 + 8x^4 + 7x^3 - 25x^3 - 200x^2 - 175x = 0$. Then determine how many of the real roots of the polynomial may be positive and how many may be negative. Factor the polynomial to confirm your results. The answer is **possible rational roots: ±1, ±5, ±7, ±25, ±35, ±175; number of possible real roots — positive: one, negative: three or one; actual roots: $x = -7, -1, 0, 5, -5$.**

After combining like terms and factoring out x, you get $x(x^4 + 8x^3 - 18x^2 - 200x - 175) = 0$. Apply the rational root theorem to the terms in the parentheses. The lead coefficient is 1, so the only rational possibilities come from the factors of the constant, ±1, ±5, ±7, ±25, ±35, and ±175.

The polynomial in the parentheses has just one sign change, or just one positive real root. When you replace each x in the parentheses with $-x$, you get $x^4 - 8x^3 - 18x^2 + 200x - 175$, and you find three sign changes (three or one negative real roots). To factor this polynomial, go back to the original format where a "convenient" arrangement allows you to factor x^3 out of each of the first three terms and $-25x$ out of the last three terms. You get $x^3(x^2 + 8x + 7) - 25x(x^2 + 8x + 7) = (x^2 + 8x + 7)(x^3 - 25x) = (x + 7)(x + 1)x(x^2 - 25) = (x + 7)(x + 1)x(x - 5)(x + 5)$. Setting this factorization equal to 0, you get roots of $-7, -1, 0, 5, -5$. Three roots are negative, one is positive, three are 0.

21 Determine all the possibilities for rational roots of the polynomial $3x^3 + 4x^2 + 30x + 40 = 0$. Then determine how many of the real roots of the polynomial may be positive and how many may be negative. Factor the polynomial to confirm your results. The answer is **possible rational roots: $\pm1, \pm2, \pm4, \pm5, \pm8, \pm10, \pm20, \pm40, \pm\frac{1}{3}, \pm\frac{2}{3}, \pm\frac{4}{3}, \pm\frac{5}{3}, \pm\frac{8}{3}, \pm\frac{10}{3}, \pm\frac{20}{3}, \pm\frac{40}{3}$;**

number of possible real roots — positive: zero, negative: three or one; actual root: $x = -\frac{4}{3}$.

Find the possible rational roots using the rational root theorem. Divide the factors of the constant ±1, ±2, ±4, ±5, ±8, ±10, ±20, and ±40 by the factors of the lead coefficient, 1 and 3.

The original polynomial has no sign changes, but replacing each x with $-x$, you get $-3x^3 + 4x^2 - 30x + 40 = 0$. The sign changes three times, so you can have three or one negative roots. Checking the roots, you factor by grouping to get $x^2(3x + 4) + 10(3x + 4) = (3x + 4)(x^2 + 10) = 0$. The only real root is $x = -\frac{4}{3}$. The second factor has no solution when you set it equal to 0.

22 Determine all the possibilities for rational roots of the polynomial $x^6 - 16x^3 + 64 = 0$. Then determine how many of the real roots of the polynomial may be positive and how many may be negative. Factor the polynomial to confirm your results. The answer is **possible rational roots: $\pm 1, \pm 2, \pm 4, \pm 8, \pm 16, \pm 32, \pm 64$; number of possible real roots — positive: two or zero, negative: zero; actual roots: $x = 2, 2$.**

Using the rational root theorem, you divide the factors of the constant, $\pm 1, \pm 2, \pm 4, \pm 8, \pm 16, \pm 32$, and ± 64, by the factors of the lead coefficient. The lead coefficient is 1, so dividing doesn't yield you any more possibilities for rational roots.

The sign changes twice in the polynomial, so you can have up to two positive roots. Replacing each x with $-x$, you get $x^6 + 16x^3 + 64$. You have no sign change here, so the roots can't be negative. The polynomial factors into $(x^3 - 8)^2 = 0$. The only solutions are 2 and 2 (two of the same 2s, too).

23 Use synthetic division to divide $(9x^4 - 37x^2 + 4) \div (x - 2)$. The answer is $9x^3 + 18x^2 - x - 2$.

$$
\begin{array}{r|rrrrr}
2 & 9 & 0 & -37 & 0 & 4 \\
  &   & 18 & 36 & -2 & -4 \\
\hline
  & 9 & 18 & -1 & -2 & 0
\end{array}
$$

The factor $x - 2$ divides evenly into the polynomial. There's no remainder.

24 Use synthetic division to divide $(2x^4 + 3x^3 + x^2 - 3x + 5) \div (x + 1)$. The answer is $2x^3 + x^2 - 3 + \dfrac{8}{x + 1}$.

Be sure to change the dividing number to -1. The remainder is 8, so you write it over the divisor.

$$
\begin{array}{r|rrrrr}
-1 & 2 & 3 & 1 & -3 & 5 \\
   &   & -2 & -1 & 0 & 3 \\
\hline
   & 2 & 1 & 0 & -3 & 8
\end{array}
$$

25 Use synthetic division to divide $(3x^3 + 4x^2 - 2x + 10) \div (x + 2)$. The answer is $3x^2 - 2x + 2 + \dfrac{6}{x + 2}$.

The remainder is 6. Write it over the divisor, $x + 2$.

$$
\begin{array}{r|rrrr}
-2 & 3 & 4 & -2 & 10 \\
   &   & -6 & 4 & -4 \\
\hline
   & 3 & -2 & 2 & 6
\end{array}
$$

26 Use synthetic division to divide $(x^4 - 12x^3 + 54x^2 - 108x + 81) \div (x - 3)$. The answer is $x^3 - 9x^2 + 27x - 27$.

The binomial divides evenly into the polynomial, so you get no remainder in this division.

$$
\begin{array}{r|rrrrr}
3 & 1 & -12 & 54 & -108 & 81 \\
  &   & 3 & -27 & 81 & -81 \\
\hline
  & 1 & -9 & 27 & -27 & 0
\end{array}
$$

27 Find $f(2)$ if $f(x) = x^5 - 3x^4 + 4x^3 - 6x^2 + 5x - 1$. The answer is $f(2) = 1$.

Use the remainder theorem and synthetic division to find the value of the function. The divisor in the synthetic division is 2; your answer is the remainder, the last element on the bottom:

$$\begin{array}{r|rrrrrr} 2 & 1 & -3 & 4 & -6 & 5 & -1 \\ & & 2 & -2 & 4 & -4 & 2 \\ \hline & 1 & -1 & 2 & -2 & 1 & 1 \end{array}$$

28 Find $f(-2)$ if $f(x) = 6x^{10} - 16x^6 + 150x^5 - 80x^2$. The answer is $f(-2) = 0$.

Use the remainder theorem and synthetic division. Don't forget to put 0s in for the missing terms in the polynomial — all the way down to the constant term.

$$\begin{array}{r|rrrrrrrrrrr} -2 & 6 & 0 & 0 & 0 & -16 & 150 & 0 & 0 & -80 & 0 & 0 \\ & & -12 & 24 & -48 & 96 & -160 & 20 & -40 & 80 & 0 & 0 \\ \hline & 6 & -12 & 24 & -48 & 80 & -10 & 20 & -40 & 0 & 0 & 0 \end{array}$$

29 If the roots of $f(x)$ are $x = 1, 2, -2, 3, 3, 3, 4, 4$, then write a function rule for $f(x)$. The answer is $f(x) = k(x - 1)(x - 2)(x + 2)(x - 3)^3(x - 4)^2$.

The root 3 appears three times, and the root 4 appears twice. Write the corresponding factors with exponents. The k represents some constant multiplier — actually, any constant multiplier except 0. The roots will be the same no matter by what constant you multiply the terms of the polynomial.

30 If the roots of $f(x)$ are $x = 0, 0, 0, 1, 1, 1, 1, -2, -3, -4$, then write a function rule for $f(x)$. The answer is $f(x) = kx^3(x - 1)^4(x + 2)(x + 3)(x + 4)$.

The root 0 appears three times, and the root 1 appears four times. Use exponents to reflect these multiple roots.

31 If the roots of $f(x)$ are $x = -1, -1, 1, \frac{1}{2}, \frac{3}{4}, -\frac{7}{8}, -\frac{7}{8}$, then write a function rule for $f(x)$. The answer is $f(x) = k(x + 1)^2(x - 1)(2x - 1)(4x - 3)(8x + 7)^2$.

The root −1 appears twice. Use the exponent to reflect this. The fractional roots have factors that use the denominator of the fraction as a coefficient of x.

Another way to look at this problem is to simplify the equations related to the fractional roots. For instance, you can simplify the equation for the root $x = \frac{3}{4}$ by multiplying each side of the equation by 4 to get $4x = 3$. Next, subtract 3 from each side to get $4x - 3 = 0$. The binomial on the left is the factor of the polynomial.

Chapter 8

Acting Rationally with Functions

A *rational number* is a number that has a fractional basis — you can write it as a fraction with an integer in both the numerator and the denominator (except that the denominator can't be 0). When the denominator of a rational number is a 1, then the number is an integer, too. For example, $\frac{4}{3}$, $\frac{10}{2}$, or $-\frac{6}{1}$ are rational numbers. The 10 divided by 2 is equal to the integer 5, and the negative 6 divided by 1 is an integer. All integers are rational numbers, but not all rational numbers are integers.

You can construct a *rational function* by putting a polynomial in both the numerator and the denominator. For example, $\frac{x^2-3}{x^3+2x^2-5x-1}$ and $\frac{4}{x^2-3x+1}$ are rational functions. As you see, a rational function may have a constant in the numerator and a polynomial in the denominator — mathematicians consider the constant in the numerator to be a polynomial with just the constant term (which technically has the variable with an exponent of 0) in it.

In this chapter, you use the properties of rational functions that arise because of their fractional nature. These functions have issues of domain and range, and the graphs can get quite interesting with *asymptotes* (lines the functions follow but generally aren't allowed to cross) here, there, and everywhere.

Determining Domain and Intercepts of Rational Functions

The *domain* of a function consists of all the real numbers that you can use as input values. You're rather selective about domain values when working with rational functions because you can't let the denominator equal 0. The number 0 comes in very handy in many cases, but it can cause problems in rational functions.

To find the intercepts of rational functions, you use the same processes as with any other function. Replace all the *x*'s with 0s to solve for the *y*-intercept; set the function rule equal to 0 (let *y* equal 0) and solve for *x* for the *x*-intercepts.

Q. Determine the domain and intercepts of the function $y = \dfrac{x^2 - 5x - 6}{x^3 - 4x}$.

A. **Domain: all real numbers except for 0, 2, and –2 (in symbols, $x \neq 0, 2, -2$); x-intercepts: (6, 0) and (–1, 0); y-intercept: none.** To determine all these answers, first factor the numerator and denominator of the function rule. You get $y = \dfrac{(x-6)(x+1)}{x(x-2)(x+2)}$. To find the domain, set the denominator equal to 0. Yes, that's generally a no-no, but this technique is a good way to solve for what you *don't* want. The denominator is 0 when x is 0, 2, or –2, so you have to eliminate those numbers from the domain. Another way to write the domain — in a more positive way than to say what it *can't* be — is to use interval notation (see Chapter 1 for details). The domain is $(-\infty, -2)$, $(-2, 0)$, $(0, 2)$, $(2, \infty)$. You can now probably see why the "everything except" wording is so popular.

To solve for the y-intercept, you usually replace all the x's with 0s. But you aren't allowed to do that in this function, because 0 isn't in the domain. That's why this function has no y-intercept. To solve for the x-intercepts, set the function rule equal to 0:

$0 = \dfrac{(x-6)(x+1)}{x(x-2)(x+2)}$. You can multiply each side of the equation by the denominator, and you have the equation $(x - 6)(x + 1) = 0$, whose solutions are $x = 6$ and –1. Those solutions are the x-values of the two x-intercepts.

1. Determine the domain and intercepts of the function $y = \dfrac{x^2 - 16}{x^2 - 3x + 2}$.

2. Determine the domain and intercepts of the function $y = \dfrac{2x^2 - 2}{(x+1)^2}$.

3. Determine the domain and intercepts of the function $y = \dfrac{x(x-3)(x+4)(x+6)}{(x^2 - 25)(x^2 - 1)}$.

4. Determine the domain and intercepts of the function $y = \dfrac{5x^3}{x^4 + 1}$.

Introducing Vertical and Horizontal Asymptotes

An asymptote is something that isn't physically there — sort of like your conscience or your make-believe playmate. An asymptote helps determine actions (much like a conscience) or the shapes of things, but it isn't really a part of the graph (or the reality of the moment).

An *asymptote* is usually a ghostlike line that helps you graph a rational function. As the curve moves along toward an asymptote, it gets closer and closer to the asymptote but never touches it. Thus, the asymptote helps determine where the graph of the function can go or not go. It gives a general shape or direction for what the graph does way out there where you can't see it.

Vertical asymptotes mark places where the function has no domain. They're vertical lines drawn in lightly or with dashes to show they're not part of the graph. You solve for the equations of the vertical asymptotes by setting the denominator of the rational function equal to 0. A curve never crosses a vertical asymptote.

Note: If the numerator and denominator have a common factor, you may have a hole in the graph rather than an asymptote. For more information, see the upcoming "Removing Discontinuities" section.

Horizontal asymptotes indicate what happens to the curve as the *x*-values get very large or very small. To find the equation of a horizontal asymptote, use the following rule:

✔ If the highest variable power in the function equation occurs in the numerator only, the function has no horizontal asymptote (though it may have an oblique asymptote — see the next section).

✔ If the highest variable power in the function equation occurs in the denominator only, the equation of the horizontal asymptote is $y = 0$ (the *x*-axis).

✔ If both the numerator and denominator have terms with the highest variable power in them, take the coefficients of those terms with the highest power and make a fraction of them. The equation of the horizontal asymptote is $y = the\ fraction\ you\ formed$. A curve can sometimes cross a horizontal asymptote when "close to the action."

Q. What are the asymptotes of the function $y = \dfrac{3x^2 + 9x - 12}{x^2 - 25}$?

A. **Vertical asymptotes: $x = 5$ and $x = -5$; horizontal asymptote: $y = 3$.** To solve for these equations, first factor the denominator and set it equal to 0: $x^2 - 25 = (x - 5)(x + 5) = 0$. The two solutions are $x = 5$ and $x = -5$. These numbers are the two values that you can't include in the domain, and the equations are the vertical asymptotes. To determine the horizontal asymptote, look at the original equation. The highest variable power is 2. You find squared terms in both the numerator and the denominator, so make a fraction of the coefficients of the x^2 terms: $\dfrac{\text{coefficient of } 3x^2}{\text{coefficient of } x^2} = \dfrac{3}{1}$. Put a y in front of the reduced equation, and you get $y = 3$.

Q. What are the asymptotes of the function $y = \dfrac{x^2 - 1}{x^4 - 81}$?

A. **Vertical asymptotes: $x = 3$ and $x = -3$; horizontal asymptote: $y = 0$.** Factor the denominator and set it equal to 0: $x^4 - 81 = (x^2 - 9)(x^2 + 9) = (x - 3)(x + 3)(x^2 + 9) = 0$. The two solutions are $x = 3$ and $x = -3$. The last factor never equals 0. You can't include the numbers 3 and -3 in the domain, so the solutions are the equations of the vertical asymptotes. To determine the horizontal asymptote, look at the original equation. The highest power is 2 in the numerator and 4 in the denominator. The highest variable power in the function rule is 4, and it occurs only in the denominator, so the equation of the horizontal asymptote is $y = 0$.

5. Determine the equations of the vertical and horizontal asymptotes of the function
$$y = \frac{x^2 - 16}{x^2 - 3x + 2}.$$

6. Determine the equations of the vertical and horizontal asymptotes of the function
$$y = \frac{9x}{x^2 - 9}.$$

7. Determine the equations of the vertical and horizontal asymptotes of the function
$$y = \frac{2x^2 - x - 3}{3x^2 - 7x - 20}.$$

8. Determine the equations of the vertical and horizontal asymptotes of the function
$$y = \frac{x^2 - 3x - 10}{x^2 - 5x - 6}.$$

Getting a New Slant with Oblique Asymptotes

An *oblique* or slant asymptote acts much like its cousins, the vertical and horizontal asymptotes. In other words, it helps you determine the ultimate direction or shape of the graph of a rational function. An oblique asymptote sometimes occurs when you have no horizontal asymptote. Oblique asymptotes take special circumstances, but the equations of these asymptotes are relatively easy to find when they do occur.

The rule for oblique asymptotes is that if the highest variable power in a rational function occurs in the numerator — and if that power is *exactly* one more than the highest power in the denominator — then the function has an oblique asymptote. You can find the equation of the oblique asymptote by dividing the numerator of the function rule by the denominator and using the first two terms in the quotient in the equation of the line that is the asymptote.

Q Find the equation of the oblique asymptote in the function $y = \frac{x^3 - 3x^2 + 3x + 1}{x^2 - 5x + 6}$.

A. **$y = x + 2$.** To find this equation, you have to divide the denominator of the function rule into the numerator. This step requires *long division*. You can't use synthetic division (which I cover in Chapter 7) because the divisor isn't a binomial in the form $x - a$. Here's what the long division looks like:

$$
\begin{array}{r}
x + 2 \\
x^2 - 5x + 6 \overline{\big)\, x^3 - 3x^2 + 3x - 1} \\
\underline{x^3 - 5x^2 + 6x} \\
2x^2 - 3x - 1 \\
\underline{2x^2 - 10x + 12} \\
7x - 13
\end{array}
$$

Ignore the remainder, and just use the first two terms in the quotient in the equation of the line.

9. Find the equation of the oblique asymptote in the function $y = \dfrac{x^2 - 5x + 4}{x - 3}$.

10. Find the equation of the oblique asymptote in the function $y = \dfrac{x^2}{x - 1}$.

11. Find the equation of the oblique asymptote in the function $y = \dfrac{x^3 - 3x^2 - 4x + 1}{x^2 - 2x + 1}$.

12. Find the equation of the oblique asymptote in the function $y = \dfrac{-3x^4 + 4x^3 + 6x^2 + 4x + 2}{x^3 + 3x^2 + 3x + 1}$.

Removing Discontinuities

A function is *discontinuous* at a point or value of x where the function can't have a domain value. Discontinuities can come in the form of gaps in huge intervals of numbers, and they can also occur at just one particular number while everything on either side of that number works. Refer to the earlier section "Determining Domain and Intercepts of Rational Functions" to see how you can have a discontinuity at any x-value that makes the denominator of a rational function equal to 0.

A *removable discontinuity,* which looks like a hole in the graph, is a bit of a misnomer. You can rewrite a function so that the function no longer appears to have a discontinuity at a particular point — in other words, you can simplify the function rule so you can't possibly get a 0 in the denominator. But the new form is just for appearances. Some functions have one or more removable discontinuities, but the discontinuity is still present, even after you've factored it out. You can't forget or ignore it. If you've factored out a discontinuity, you have to limit the domain of the new function rule ($x \neq ?$). The changed form in the function is just for convenience in evaluating and graphing.

If $y = \dfrac{f(x)}{g(x)}$ is equal to $\dfrac{0}{0}$ for some value of $x = c$, and if $(x - c)$ is a factor of both $f(x)$ and $g(x)$, then c is a *removable discontinuity* of the rational function.

Q. Determine where the function

$$y = \frac{x^2 - 4}{x^2 + 4x + 4}$$

has a removable discontinuity. Remove the discontinuity and rewrite the function rule.

A. **Removable discontinuity: $x = -2$; revised function rule: $y = \frac{x-2}{x+2}, x \neq -2$.** If you let $x = -2$ in the original equation, you get

$$y = \frac{(-2)^2 - 4}{(-2)^2 + 4(-2) + 4} = \frac{4 - 4}{4 - 8 + 4} = \frac{0}{0}.$$

Factor the numerator and denominator to get $y = \frac{(x-2)(x+2)}{(x+2)(x+2)}$. The factor $(x + 2)$ occurs in both, so you find a removable discontinuity at $x = -2$. Divide the numerator and denominator by the common factor to get $y = \frac{(x-2)(x+2)}{(x+2)(x+2)} = \frac{x-2}{x+2}$. You can write the original function as $y = \frac{x-2}{x+2}, x \neq -2$.

13. Determine where the function $y = \frac{x^2 - 7x - 18}{x^2 - 17x + 72}$ has a removable discontinuity. Remove the discontinuity and rewrite the function rule.

14. Determine where the function $y = \frac{x^3 - 3x^2 + 2x}{x^3 - 4x}$ has a removable discontinuity. Remove the discontinuity and rewrite the function rule.

15. Determine where the function $y = \frac{x^3 + 3x^2 - 4x - 12}{x^3 + 3x^2 - x - 3}$ has a removable discontinuity. Remove the discontinuity and rewrite the function rule.

16. Determine where the function $y = \frac{x^3 - 3x^2 - 16x - 12}{x^3 - 4x^2 - 11x - 6}$ has a removable discontinuity. Remove the discontinuity and rewrite the function rule.

Going the Limit: Limits at a Number and Infinity

Finding limits, which is especially important in calculus, is all about testing boundaries. The *limit* of a function is a y-value that the function approaches from both sides of a particular x-value or that y approaches as x gets very large or very small. The x-value may or may not be in the domain of the function, but you can have values in the domain on either side of it (really close to it).

To find the limit of $f(x)$ at the x-value a, replace all the x's in the function rule with a and simplify. If you get a fraction with 0's in the numerator and denominator, remove the discontinuity, if possible (see the preceding section called "Removing Discontinuities") and evaluate $f(a)$ again. The notation for finding a limit at a is $\lim_{x \to a} f(x) = L$, where L is the y-value or limit.

To find the limit at infinity, replace all the x's with ∞ and simplify. If you get a fraction with ∞ in both the numerator and denominator, then divide each term in the fraction by the highest variable power and evaluate again.

The key to solving the limits of infinity is usually to recall that the limit of a constant divided by infinity is 0, $\lim_{x\to\infty} \frac{c}{x} = 0$.

Q. Find the limit of $f(x) = \dfrac{x^2 - 5x + 4}{2x^2 - 11x + 12}$ at $x = 4$.

A. $\dfrac{3}{5}$. Replacing each x with 4 results in a fraction with 0s in both the numerator and denominator:

$$\lim_{x\to4} \frac{x^2 - 5x + 4}{2x^2 - 11x + 12} - \frac{4^2 - 5(4) + 4}{2(4)^2 - 11(4) + 12}$$

$$- \frac{16 - 20 + 4}{32 - 44 + 12} - \frac{0}{0}$$

Factor and reduce to get

$$\lim_{x\to4} \frac{x^2 - 5x + 4}{2x^2 - 11x + 12} = \lim_{x\to4} \frac{(x-4)(x-1)}{(2x-3)(x-4)}$$

$$= \lim_{x\to4} \frac{x-1}{2x-3} = \frac{4-1}{8-3} = \frac{3}{5}$$

Even though the function is undefined at $x = 4$, it has a limit of $\frac{3}{5}$, which means that the function values get really close to that limit or y-value when you use x's slightly smaller and slightly larger than 4.

Q. Find the limit of $f(x) = \dfrac{3x}{x-4}$ at $x = 4$

A. **No limit.** The function has no limit at 4: $\lim_{x\to4} \frac{3x}{x-4} = \frac{3(4)}{4-4} = \frac{12}{0}$. You can't reduce the function rule. You can get a 0 in the denominator but not in the numerator.

Q. Find the limit of $f(x) = \dfrac{6x^2 - 2x + 3}{2x^2 + 5x - 2}$ as $x \to \infty$ by dividing each term by x^2

A. **3.** When you plug in ∞ for x, you get

$$\lim_{x\to\infty} \frac{6x^2 - 2x + 3}{2x^2 + 5x - 2} = \frac{6(\infty)^2 - 2\infty + 3}{2(\infty)^2 + 5\infty - 2} = \frac{\infty}{\infty}.$$

A fraction with infinity over infinity doesn't tell you a thing — just that something big is dividing something else big. You need to be able to compare these huge numbers, so divide each term in the function by x^2 and evaluate again:

$$\lim_{x\to\infty} \frac{\frac{6x^2}{x^2} - \frac{2x}{x^2} + \frac{3}{x^2}}{\frac{2x^2}{x^2} + \frac{5x}{x^2} - \frac{2}{x^2}} = \lim_{x\to\infty} \frac{6 - \frac{2}{x} + \frac{3}{x^2}}{2 + \frac{5}{x} - \frac{2}{x^2}} =$$

$$\frac{6 - \frac{2}{\infty} + \frac{3}{\infty^2}}{2 + \frac{5}{\infty} - \frac{2}{\infty^2}} = \frac{6 - 0 + 0}{2 + 0 - 0} = \frac{6}{2} = 3.$$

Whenever you divide a constant by infinity, that fraction equals 0; your limit for this function is 3.

Q. Find the limit of $f(x) = \dfrac{4x^2 - 11x + 2}{2x + 47}$ as $x \to \infty$.

A. **No limit.** This function has no limit as x approaches infinity:

$$\lim_{x\to\infty} \frac{4x^2 - 11x + 2}{2x + 47} = \lim_{x\to\infty} \frac{\frac{4x^2}{x^2} - \frac{11x}{x^2} + \frac{2}{x^2}}{\frac{2x}{x^2} + \frac{47}{x^2}}$$

$$= \lim_{x\to\infty} \frac{4 - \frac{11}{x} + \frac{2}{x^2}}{\frac{2}{x} + \frac{47}{x^2}} = \frac{4 - \frac{11}{\infty} + \frac{2}{\infty^2}}{\frac{2}{\infty} + \frac{47}{\infty^2}} = \frac{4 - 0 + 0}{0 + 0} = \frac{4}{0}$$

You can't divide by 0.

17. Find the limit of $f(x) = \dfrac{x^2+5}{x^2-2}$ at $x = 3$.

18. Find the limit of $f(x) = \dfrac{3x^2-5x-2}{x^2-3x+2}$ at $x = 2$.

19. Find the limit of $f(x) = \dfrac{15x^2+6x+9}{3x^2+7x+11}$ as $x \to \infty$.

20. Find the limit of $f(x) = \dfrac{4x^2+3x-1}{2x^3+9x+11}$ as $x \to \infty$.

21. Find the limit of $f(x) = \dfrac{x^2-1}{x^2-2x+1}$ at $x = 1$.

22. Find the limit of $f(x) = \dfrac{2-8x^2-5x^3}{4x^2+3x+1}$ as $x \to \infty$.

Graphing Rational Functions

A rational function is a polynomial divided by another polynomial. The rational function rule makes the separate parts (the polynomials) of the graph pretty much smooth curves. What distinguishes the graphs of rational functions from those of polynomial functions is that rational functions have breaks or gaps or discontinuities shown with vertical asymptotes, as well as horizontal and oblique asymptotes that show other patterns of behavior.

To graph a rational function, follow these steps:

1. Determine all intercepts.

2. Determine the domain and, at the same time, the equations of the vertical asymptotes.

3. Determine the equations of any horizontal or oblique asymptotes.

4. Sketch in all the intercepts and asymptotes.

5. Solve for any other additional points you need to finish drawing the graph.

EXAMPLE

Q. Sketch the graph of the function $y = \dfrac{x^2 - 3x - 10}{x^2 - 9}$, indicating all intercepts and asymptotes.

A. **The y-intercept is $\left(0, \dfrac{10}{9}\right)$, and the x-intercepts are $(5, 0)$ and $(-2, 0)$.** The domain consists of all real numbers except for $x = 3$ and $x = -3$, which are the equations of the vertical asymptotes. The equation of the horizontal asymptote is $y = 1$. Figure 8-1a shows the intercepts and asymptotes sketched in. As you can see, these points alone aren't too much help with the graph. But here are some observations you can make: The graph of the function to the left of $x = -3$ has to be above the horizontal asymptote, because in that interval, no x-intercepts allow the curve to cross the axis. The graph to the right of $x = 3$ has to be below the horizontal asymptote, crossing the x-axis at the intercept. And the curve between the vertical asymptotes probably rises from left to right, crossing the horizontal asymptote but hugging the two vertical asymptotes. Substituting some other values of x into the function rule, additional points include $(-9, 1.361)$, $(-5, 1.875)$, $(-2.8, -5.38)$, $(1, 1.5)$, $(2, 2.4)$, $(4, -0.857)$, and $(7, 0.45)$, which you can use to complete the sketch of the function, shown in Figure 8-1b.

Q. Sketch the graph of the function $y = \dfrac{x^2 - 3x - 4}{x + 2}$, indicating all intercepts and asymptotes.

A. **The y-intercept is $(0, -2)$, and the x-intercepts are $(4, 0)$ and $(-1, 0)$.** You have a vertical asymptote of $x = -2$ and an oblique asymptote of $y = x - 5$. Figure 8-2a shows those points and asymptotes sketched in. From these points, you get a sense of what's happening in the upper-right area, but you need more help with the rest of the graph. You can choose additional points to help with the rest of the sketch, like $(-8, -14)$, $(-4, -12)$, and $(-3, -14)$, shown in Figure 8-2b.

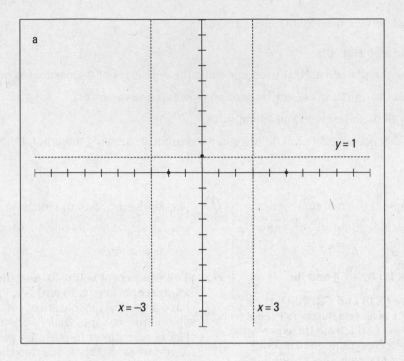

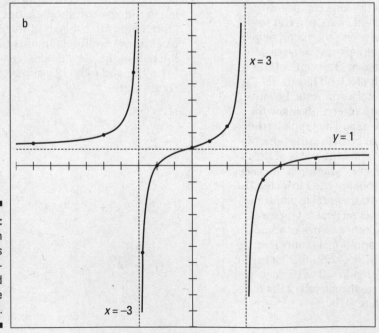

Figure 8-1:
Start with
some points
and asymp-
totes and
fill in the
details.

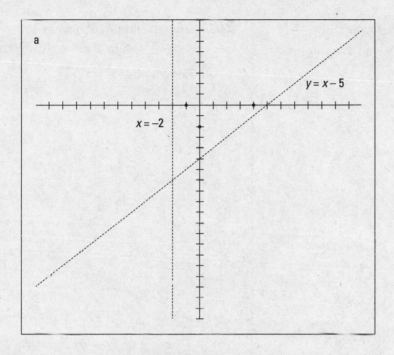

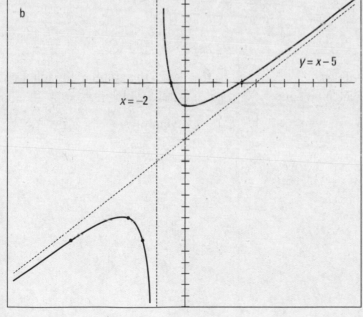

Figure 8-2:
The vertical
and slant
asymptotes
give shape
to the curve.

23. Sketch the graph of the function $y = \frac{x+3}{x-1}$, indicating all intercepts and asymptotes.

24. Sketch the graph of the function $y = \frac{x}{x^2 - 5x - 6}$, indicating all intercepts and asymptotes.

25. Sketch the graph of the function $y = \frac{x^2 - 16}{x^2 + 2x - 15}$, indicating all intercepts and asymptotes.

26. Sketch the graph of the function $y = \frac{x^2 - 5x + 4}{x - 3}$, indicating all intercepts and asymptotes.

Answers to Problems on Rational Functions

The following are the answers to the practice problems presented earlier in this chapter.

1 Determine the domain and intercepts of the function $y = \dfrac{x^2 - 16}{x^2 - 3x + 2}$. The answer is **domain: $(-\infty, 1)$, $(1, 2)$, $(2, \infty)$; intercepts: $(0, -8)$, $(4, 0)$, $(-4, 0)$.**

The denominator factors into $(x - 1)(x - 2)$. Therefore, the domain consists of all the real numbers except for 1 and 2. Those values make the denominator equal 0. You can find the y-intercept by letting all the x's in the function rule equal 0 and simplifying the fraction. Determine the x-intercepts by setting the function equation equal to 0, multiplying each side of the equation by the denominator, and then solving the quadratic equation that results.

2 Determine the domain and intercepts of the function $y = \dfrac{2x^2 - 2}{(x + 1)^2}$. The answer is **domain: $(-\infty, -1)$, $(-1, \infty)$; intercepts: $(0, -2)$, $(1, 0)$.**

The domain contains all real numbers except for the number -1, because -1 makes the denominator equal to 0. Calculate the y-intercept by letting all the x's in the function rule equal 0 and solving for y. You can find the x-intercept *after* factoring and reducing the function rule. The factor $x + 1$ divides out, leaving $2(x - 1)$ in the numerator.

3 Determine the domain and intercepts of the function $y = \dfrac{x(x - 3)(x + 4)(x + 6)}{(x^2 - 25)(x^2 - 1)}$. The answer is **domain: $(-\infty, -5)$, $(-5, -1)$, $(-1, 1)$, $(1, 5)$, $(5, \infty)$; intercepts: $(0, 0)$, $(3, 0)$, $(-4, 0)$, $(-6, 0)$.**

The domain consists of all real numbers except for -5, -1, 1, and 5. The y-intercept is also an x-intercept at $(0, 0)$. Find the other x-intercepts by setting the function equation equal to 0 and solving for x.

4 Determine the domain and intercepts of the function $y = \dfrac{5x^3}{x^4 + 1}$. The answer is **domain: all real numbers; intercept: $(0, 0)$.**

The domain contains all real numbers. No x-value can make the denominator equal to 0, so you can use all x's. When $x = 0$, $y = 0$. The origin is the only intercept.

5 Determine the equations of the vertical and horizontal asymptotes of the function $y = \dfrac{x^2 - 16}{x^2 - 3x + 2}$. The answer is **vertical: $x = 1$, $x = 2$; horizontal: $y = 1$.**

Determine the vertical asymptotes by setting the denominator equal to 0 and solving for x. Find the horizontal asymptote by making a fraction of the coefficients of the x^2 terms and setting the result equal to y.

6 Determine the equations of the vertical and horizontal asymptotes of the function $y = \dfrac{9x}{x^2 - 9}$. The answer is **vertical: $x = 3$, $x = -3$; horizontal: $y = 0$.**

Find the vertical asymptotes by setting the denominator equal to 0 and solving for x. The horizontal asymptote is $y = 0$, because the highest power of the variable is in the denominator of the fraction.

7 Determine the equations of the vertical and horizontal asymptotes of the function $y = \dfrac{2x^2 - x - 3}{3x^2 - 7x - 20}$. The answer is **vertical: $x = 4$, $x = -\dfrac{5}{3}$; horizontal: $y = \dfrac{2}{3}$.**

You can get the vertical asymptotes by setting the denominator equal to 0 and solving for x: $0 = 3x^2 - 7x - 20 = (3x + 5)(x - 4)$; $x = -\frac{5}{3}, 4$. For the horizontal asymptote, make a fraction of the coefficients of the x^2 terms and set the result equal to y.

8 Determine the equations of the vertical and horizontal asymptotes of the function $y = \frac{x^2 - 3x - 10}{x^2 - 5x - 6}$. The answer is **vertical: $x = 6$, $x = -1$; horizontal: $y = 1$.**

Find the vertical asymptotes by setting the denominator equal to 0 and solving for x. Determine the horizontal asymptote by making a fraction of the coefficients of the x^2 terms and setting the result equal to y.

9 Find the equation of the oblique asymptote in the function $y = \frac{x^2 - 5x + 4}{x - 3}$. The answer is $y = x - 2$.

Use synthetic division or long division to divide the denominator into the numerator:

$$
\begin{array}{r|rrr}
3 & 1 & -5 & 4 \\
& & 3 & -6 \\
\hline
& 1 & -2 & -2
\end{array}
$$

The first two terms in the quotient are the slope and y-intercept of the oblique asymptote's equation.

10 Find the equation of the oblique asymptote in the function $y = \frac{x^2}{x - 1}$. The answer is $y = x + 1$.

Use synthetic division or long division to divide the denominator into the numerator:

$$
\begin{array}{r|rrr}
1 & 0 & 0 \\
& & 1 & 1 \\
\hline
& 1 & 1 & 1
\end{array}
$$

The first two terms in the quotient are the slope and y-intercept of the oblique asymptote's equation.

11 Find the equation of the oblique asymptote in the function $y = \frac{x^3 - 3x^2 - 4x + 1}{x^2 - 2x + 1}$. The answer is $y = x - 1$.

Use long division to divide the denominator into the numerator:

$$
\begin{array}{r}
x - 1 \\
x^2 - 2x + 1 \overline{\smash{\big)}\, x^3 - 3x^2 - 4x + 1} \\
\underline{x^3 - 2x^2 + x} \\
-x^2 - 5x + 1 \\
\underline{-x^2 + 2x - 1} \\
-7x + 2
\end{array}
$$

The first two terms in the quotient are the slope and y-intercept of the oblique asymptote's equation.

12 Find the equation of the oblique asymptote in the function $y = \frac{-3x^4 + 4x^3 + 6x^2 + 4x + 2}{x^3 + 3x^2 + 3x + 1}$. The answer is $y = -3x + 13$.

Use long division to divide the denominator into the numerator:

$$
\begin{array}{r}
-3x+13 \\
x^3+3x^2+3x+1\overline{\smash{\big)}\,-3x^4+4x^3+6x^2+4x+2} \\
\underline{-3x^4-9x^3-9x^2-3x} \\
13x^3+15x^2+7x+2 \\
\underline{13x^3+39x^2+39x+13} \\
-24x^2-32x-11
\end{array}
$$

The first two terms in the quotient are the slope and y-intercept of the oblique asymptote's equation.

13 Determine where the function $y=\dfrac{x^2-7x-18}{x^2-17x+72}$ has a removable discontinuity. Remove the discontinuity and rewrite the function rule. The answer is **removable discontinuity: $x = 9$; revised function rule: $y=\dfrac{x+2}{x-8}$, $x \neq 9$.**

The function rule factors, and you have a common factor of $x - 9$ in the numerator and denominator. Divide through by $x - 9$: $y=\dfrac{x^2-7x-18}{x^2-17x+72}=\dfrac{\cancel{(x-9)}(x+2)}{(x-8)\cancel{(x-9)}}$.

14 Determine where the function $y=\dfrac{x^3-3x^2+2x}{x^3-4x}$ has a removable discontinuity. Remove the discontinuity and rewrite the function rule. The answer is **removable discontinuities: $x = 0, 2$; revised function rule: $y=\dfrac{x-1}{x+2}$, $x \neq 0,2$.**

The function rule factors, and you find two common factors: x and $x - 2$. Divide through by both common factors: $y=\dfrac{x^3-3x^2+2x}{x^3-4x}=\dfrac{\cancel{x}(x-1)\cancel{(x-2)}}{\cancel{x}\cancel{(x-2)}(x+2)}$.

15 Determine where the function $y=\dfrac{x^3+3x^2-4x-12}{x^3+3x^2-x-3}$ has a removable discontinuity. Remove the discontinuity and rewrite the function rule. The answer is **removable discontinuity: $x = -3$; revised function rule: $y=\dfrac{x^2-4}{x^2-1}$, $x \neq -3$.**

The function rule factors, and you have a common factor of $x + 3$ in the numerator and denominator. Divide through by $x + 3$: $y=\dfrac{x^3+3x^2-4x-12}{x^3+3x^2-x-3}=\dfrac{x^2(x+3)-4(x+3)}{x^2(x+3)-1(x+3)}=\dfrac{\cancel{(x+3)}(x^2-4)}{\cancel{(x+3)}(x^2-1)}=\dfrac{x^2-4}{x^2-1}$.

16 Determine where the function $y=\dfrac{x^3-3x^2-16x-12}{x^3-4x^2-11x-6}$ has a removable discontinuity. Remove the discontinuity and rewrite the function rule. The answer is **removable discontinuities: $x = -1, 6$; revised function rule: $y=\dfrac{x+2}{x+1}$, $x \neq -1,6$.**

The function rule factors, and the numerator and denominator have two common factors: $x + 1$ and $x - 6$. Divide through by both common factors: $y=\dfrac{x^3-3x^2-16x-12}{x^3-4x^2-11x-6}=\dfrac{\cancel{(x+1)}(x+2)\cancel{(x-6)}}{\cancel{(x+1)}(x+1)\cancel{(x-6)}}=\dfrac{x+2}{x+1}$.

17 Find the limit of $f(x) = \dfrac{x^2 + 5}{x^2 - 2}$ at $x = 3$. The answer is **2.**

Replace each x with 3 to get $\displaystyle\lim_{x \to 3} \frac{x^2 + 5}{x^2 - 2} = \frac{9 + 5}{9 - 2} = \frac{14}{7} = 2$.

18 Find the limit of $f(x) = \dfrac{3x^2 - 5x - 2}{x^2 - 3x + 2}$ at $x = 2$. The answer is **7.**

Replacing each x with 2 gives you a 0 in both numerator and denominator. Factor the numerator and denominator, divide out the common factor, and substitute the 2 into the new form:

$$\lim_{x \to 2} \frac{3x^2 - 5x - 2}{x^2 - 3x + 2} = \lim_{x \to 2} \frac{(3x + 1)(x - 2)}{(x - 2)(x - 1)} = \lim_{x \to 2} \frac{3x + 1}{x - 1} = \frac{6 + 1}{2 - 1} = \frac{7}{1} = 7$$

19 Find the limit of $f(x) = \dfrac{15x^2 + 6x + 9}{3x^2 + 7x + 11}$ as $x \to \infty$. The answer is **5.**

Divide each term by x^2 and then replace each x with ∞:

$$\lim_{x \to \infty} \frac{15x^2 + 6x + 9}{3x^2 + 7x + 11} = \lim_{x \to \infty} \frac{\dfrac{15x}{x^2} + \dfrac{6x}{x^2} + \dfrac{9}{x^2}}{\dfrac{3x^2}{x^2} + \dfrac{7x}{x^2} + \dfrac{11}{x^2}} = \lim_{x \to \infty} \frac{15 + \dfrac{6}{x} + \dfrac{9}{x^2}}{3 + \dfrac{7}{x} + \dfrac{11}{x^2}}$$

$$= \frac{15 + \dfrac{6}{\infty} + \dfrac{9}{\infty^2}}{3 + \dfrac{7}{\infty} + \dfrac{11}{\infty^2}} = \frac{15 + 0 + 0}{3 + 0 + 0} = \frac{15}{3} = 5$$

20 Find the limit of $f(x) = \dfrac{4x^2 + 3x - 1}{2x^3 + 9x + 11}$ as $x \to \infty$. The answer is **0.**

Divide each term by x^3, and then replace each x with ∞:

$$\lim_{x \to \infty} \frac{15x^2 + 6x + 9}{3x^2 + 7x + 11} = \lim_{x \to \infty} \frac{\dfrac{15x}{x^2} + \dfrac{6x}{x^2} + \dfrac{9}{x^2}}{\dfrac{3x^2}{x^2} + \dfrac{7x}{x^2} + \dfrac{11}{x^2}} = \lim_{x \to \infty} \frac{15 + \dfrac{6}{x} + \dfrac{9}{x^2}}{3 + \dfrac{7}{x} + \dfrac{11}{x^2}}$$

$$= \frac{15 + \dfrac{6}{\infty} + \dfrac{9}{\infty^2}}{3 + \dfrac{7}{\infty} + \dfrac{11}{\infty^2}} = \frac{15 + 0 + 0}{3 + 0 + 0} = \frac{15}{3} = 5$$

21 Find the limit of $f(x) = \dfrac{x^2 - 1}{x^2 - 2x + 1}$ at $x = 1$. **No limit.**

Even after factoring the numerator and denominator and reducing the fraction, you still get a 0 in the denominator when replacing the x's with 1. This function has no limit at $x = 1$:

$$\lim_{x \to 1} \frac{x^2 - 1}{x^2 - 2x + 1} = \lim_{x \to 1} \frac{(x - 1)(x + 1)}{(x - 1)(x - 1)} = \lim_{x \to 1} \frac{x + 1}{x - 1} = \frac{2}{0}$$

22 Find the limit of $f(x) = \dfrac{2 - 8x^2 - 5x^3}{4x^2 + 3x + 1}$ as $x \to \infty$. **No limit.**

After dividing each term by x^3 and replacing each x with ∞, you end up with a fraction that has 0 in the denominator. This function has no limit as x approaches infinity:

$$\lim_{x \to \infty} \frac{15x^2 + 6x + 9}{3x^2 + 7x + 11} = \lim_{x \to \infty} \frac{\dfrac{15x^2}{x^2} + \dfrac{6x}{x^2} + \dfrac{9}{x^2}}{\dfrac{3x^2}{x^2} + \dfrac{7x}{x^2} + \dfrac{11}{x^2}} = \lim_{x \to \infty} \frac{15 + \dfrac{6}{x} + \dfrac{9}{x^2}}{3 + \dfrac{7}{x} + \dfrac{11}{x^2}}$$

$$= \frac{15 + \dfrac{6}{\infty} + \dfrac{9}{\infty^2}}{3 + \dfrac{7}{\infty} + \dfrac{11}{\infty^2}} = \frac{15 + 0 + 0}{3 + 0 + 0} = \frac{15}{3} = 5$$

23 Sketch the graph of the function $y = \dfrac{x+3}{x-1}$, indicating all intercepts and asymptotes.

The intercepts are $(0, -3)$ and $(-3, 0)$. The vertical asymptote is $x = 1$, and the horizontal asymptote is $y = 1$.

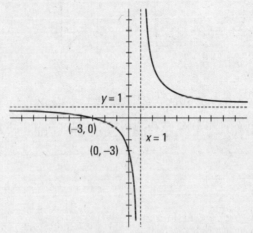

24 Sketch the graph of the function $y = \dfrac{x}{x^2 - 5x - 6}$, indicating all intercepts and asymptotes.

The only intercept is $(0, 0)$. The vertical asymptotes are $x = 6$ and $x = -1$, and the horizontal asymptote is $y = 0$.

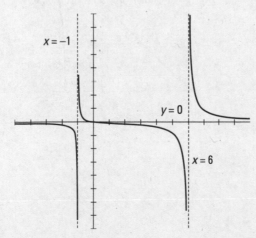

25 Sketch the graph of the function $y = \dfrac{x^2 - 16}{x^2 + 2x - 15}$, indicating all intercepts and asymptotes.

The intercepts are $\left(0, \dfrac{16}{15}\right)$, $(4, 0)$, and $(-4, 0)$. The vertical asymptotes are $x = -5$ and $x = 3$, and the horizontal asymptote is $y = 1$.

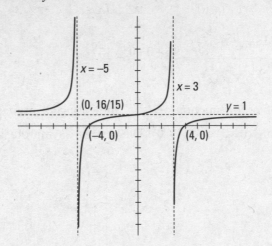

26 Sketch the graph of the function $y = \dfrac{x^2 - 5x + 4}{x - 3}$, indicating all intercepts and asymptotes.

The intercepts are $\left(0, -\dfrac{4}{3}\right)$, $(4, 0)$, and $(1, 0)$. The vertical asymptote is $x = 3$, and the oblique asymptote is $y = x - 2$.

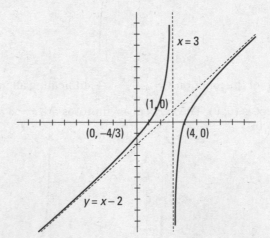

Chapter 9

Exposing Exponential and Logarithmic Functions

· ·

In This Chapter

▶ Introducing the number e

▶ Working with properties of exponential and logarithmic expressions

▶ Solving exponential and logarithmic equations

▶ Money matters: Using compound interest formulas

▶ Graphing exponential and logarithmic functions

· ·

Exponential and logarithmic functions are related to one another, but they're quite different from the algebraic functions such as polynomials, radical functions, and rational functions. Algebraic functions all use the variable as the base of action. Exponential functions have the variable in the exponent (hence, the name), and logarithmic functions have the variable in the argument (there'll be no disagreement there — I explain arguments later in this chapter)!

Here, you practice using these new functions and simplify expressions that have exponential and logarithmic terms and factors in them. Solving equations of exponential and logarithmic functions takes some different types of twists and turns. Lastly, you get to graph exponential and logarithmic functions, which requires knowing where and how the C-shaped curve comes from the function equation. Read on.

Evaluating e-Expressions and Powers of e

When mathematicians and scientists assigned the letter e to represent the number 2.71828 . . . , little did they know that there'd be e-mail, eBay, eMall, and so on. Look what they started! In the world of mathematics, the letter e is a constant value — slightly smaller than the number 3. The more frequently used exponential and logarithmic functions are based on the constant e, although people use other bases on occasion. The exponential rules apply for all the constant bases.

Here's a quick review of the rules you need:

✔ $b^x \cdot b^y = b^{x+y}$

✔ $\dfrac{b^x}{b^y} = b^{x-y}, b \neq 0$

✔ $(b^x)^y = b^{xy}$

✔ $b^0 = 1$

Q. Given the exponential function $f(x) = 5^{x+1}$, determine $f(-1)$, $f(0)$, and $f(2)$.

A. $f(-1) = 1; f(0) = 5; f(2) = 125$.
Substituting the input value for x in each case, $f(-1) = 5^{-1+1} = 5^0 = 1$, $f(0) = 5^{0+1} = 5^1 = 5$, and $f(2) = 5^{2+1} = 5^3 = 125$.

Q. Given the exponential function $f(x) = xe^{x+1}$, determine $f(-1)$, $f(0)$, and $f(2)$.

A. $f(-1) = -1; f(0) = 0; f(2) = 2e^3$.
Substituting the input value for x in each case, $f(-1) = -1e^{-1+1} = -1 \cdot e^0 = -1 \cdot 1 = -1$, $f(0) = 0 \cdot e^{0+1} = 0 \cdot e^1 = 0$, and $f(2) = 2 \cdot e^{2+1} = 2e^3$. You usually leave powers of e just as they are; don't bother finding a decimal approximation.

1. Given the exponential function $f(x) = 2^{x-1}$, determine $f(0)$, $f(1)$, and $f(2)$.

2. Given the exponential function $f(x) = 3^x(1 - 2 \cdot 3^x)$, determine $f(0)$, $f(1)$, and $f(2)$.

3. Given the exponential function $f(x) = e^x - 2e^{2x}$, determine $f(0)$ and $f(1)$.

4. Given the exponential function $f(x) = (e^x + 2)^2 - (e^x - 2)^2$, determine $f(0)$, $f(1)$, and $f(2)$.

5. Given the exponential function $f(x) = 2(e^{x+1})^2$, determine $f(0)$, $f(1)$, and $f(-1)$.

6. Given the exponential function $f(x) = x^e - e^x$, determine $f(0)$, $f(1)$, and $f(2)$.

Solving Exponential Equations

In the real world, people use exponential equations to show how something grows or decays. These little wonders can help you figure out how long radiation is going to stick around, estimate the age of an artifact from an archeological dig, or determine how many bacteria will be teeming in a petri dish in a couple hours. Or, as the next section shows you, they can help you track the cash in your bank account.

An *exponential equation* has an exponential expression in it. In an exponential equation, you may have 2^x or e^x or some other base with a variable as its exponent. In general, to solve an exponential equation, you want to adjust your equation so you can set two numbers that have the same base equal to one another, or you want to factor a quadratic-like trinomial (see Chapter 2) or some binomial so you can apply the multiplication property of zero.

Q. Solve for x in $4^{2x+1} = 8^{3x+4}$.

A. $x = -2$. First change the 4 and 8 to powers of 2 so your bases match: $(2^2)^{2x+1} = (2^3)^{3x+4}$. Next, using the property of exponents that lets you raise a power to a power, you can write $2^{4x+2} = 2^{9x+12}$. Set the two exponents equal to one another, and solve the linear equation: $4x + 2 = 9x + 12$; $-5x = 10$; $x = -2$. In the original equation, if you replace the x with -2, you get $4^{-3} = 8^{-2}$, or $\frac{1}{4^3} = \frac{1}{8^2}$, $\frac{1}{64} = \frac{1}{64}$.

Q. Solve for x in $3^{2x} - 12(3^x) + 27 = 0$.

A. $x = 1$ or $x = 2$. This equation is quadratic-like. Replace each 3^x with y and think of the equation as having the format $y^2 - 12y + 27 = (y - 3)(y - 9) = 0$. Then you can factor the original equation as $(3^x - 3)(3^x - 9) = 0$. Using the multiplication property of zero, you set each factor equal to zero and solve for x. When $3^x - 3 = 0$, $3^x = 3^1$, giving you $x = 1$. When $3^x - 9 = 0$, $3^x = 9 = 3^2$, so $x = 2$.

7. Solve for x in $2^x = 4^{x+1}$.

8. Solve for x in $9^{x-5} = \dfrac{1}{27^x}$.

9. Solve for x in $e^{4x-2} = e^{x+7}$.

10. Solve for x in $2^{2x} - 5 \cdot 2^x + 4 = 0$.

11. Solve for x in $e^x - 1 = 0$.

12. Solve for x in $xe^x + 5e^x = 0$.

Making Cents: Applying Compound Interest and Continuous Compounding

When you deposit your inheritance in an account that earns 4.5 percent interest, compounded quarterly, your money then earns interest on the deposit, and the interest earns interest, and the interest on the interest earns interest — well, I'm sure you get the point here. Compound interest is a very powerful instrument and the basis of the saving and borrowing that you do in your life.

The formula for determining the total amount of money you accumulate in an account after t years is $A = P\left(1 + \dfrac{r}{n}\right)^{nt}$, where A is the total amount of money accumulated, P is the principal (initial deposit), r is the rate of interest (written as a decimal), and n is the number of times each year that the interest is computed.

The formula for determining the total amount of money accumulated when it's compounded *continuously* is $A = Pe^{rt}$. People use the continuously compounding formula at times for convenience of figuring and also in other financial formulas that deal with the present and future value of an investment. Essentially, in continuous compounding, the n, or number of times you compound, is infinity — every little part of every second.

Q. Determine the total amount of money accumulated after 10 years if you deposit your inheritance of $50,000 in an account that earns 4.5% interest, compounded quarterly.

A. **$78,218.84.** Using the compound interest formula, you get
$$A = 50,000\left(1 + \frac{0.045}{4}\right)^{4(10)}$$
$$= 50,000(1.01125)^{40} \approx 78,218.84.$$
Your money grows to almost $80,000.

Q. Determine the total amount of money accumulated after 10 years if you deposit your inheritance of $50,000 in an account that earns 4.5% interest, compounded continuously.

A. **$78,415.61.** Using the compound interest formula, you get $A = 50,000e^{0.045(10)} = 50,000e^{0.45} \approx 78,415.61$. Your money grows to almost $80,000, just as it does with the quarterly compounding. The difference is less than $200.

13. Determine the total amount of money accumulated after 20 years if you deposit $1,000,000 in an account that earns 5.25% interest, compounded monthly.

14. Determine the total amount of money accumulated after 2 years and 6 months if you deposit $25,000 in an account that earns 5% interest, compounded daily (use 365 days).

15. Determine the total amount of money accumulated after 20 years if you deposit $1,000,000 in an account that earns 5.25% interest, compounded continuously.

16. Determine the total amount of money accumulated after 2 years and 6 months if you deposit $25,000 in an account that earns 5% interest, compounded continuously.

Checking Out the Properties of Logarithms

With logarithmic functions, numbers rise or fall quickly and then level out. Mathematicians, scientists, and others use logs to determine just how bad that earthquake was, just how loud that music coming from the car next to you is, and just how long ago that person became dearly departed. Logarithms work for all these situations and more. They aren't easy to use, so calculators or charts are necessary, but logarithms tell it like it is — or was.

A logarithmic function is the inverse of some exponential function. Here's the relationship between a logarithmic function with base b and an exponential function with base b:

$$y = \log_b x \leftrightarrow b^y = x$$

Mathematicians say that the log function $f(x) = \log_b x$ has a *base* of b and an *argument* x. When you see $\log_b x$, the output of the function is an exponent: This formula is saying, "The answer is the exponent that you put on b to get the number x." Logarithms have some very handy properties — all linked to their respective exponential partners.

The most frequently used logarithmic bases are base 10 and base e. Base 10 logarithms are called the *common* logs, and base e logarithms are the *natural* logs. If you have a scientific or graphing calculator, you probably have a *log* button and an *ln* button. No subscripts are shown — they're just understood to be included. The *log* button is shorthand for base 10 or common logs, and the *ln* button is for the natural logs. I show the following rules for logs with base *b* (any base, including 10) and for natural logs.

Laws of Logarithms	*For Log Base b*	*For Natural Log*
Log of a product	$\log_b xy = \log_b x + \log_b y$	$\ln xy = \ln x + \ln y$
Log of a quotient	$\log_b \frac{x}{y} = \log_b x - \log_b y$	$\ln \frac{x}{y} = \ln x - \ln y$
Log of a power	$\log_b x^n = n\log_b x$	$\ln x^n = n\ln x$
Log of 1	$\log_b 1 = 0$	$\ln 1 = 0$
Log of the base	$\log_b b = 1$	$\ln e = 1$
Log of the reciprocal	$\log_b \frac{1}{y} = -\log_b y$	$\ln \frac{1}{y} = -\ln y$
Change of bases	$\log_b x = \frac{\log x}{\log b} = \frac{\ln x}{\ln b}$	

Note: You don't need the change of bases rule for the problems in this book, but this rule can really come in handy if you want to use your calculator for logarithms with bases other than 10 or *e*.

People use the various logarithm rules to make the logarithmic expressions simpler and easier to handle when solving equations, doing higher mathematics, and using some scientific applications.

Q. Simplify the logarithmic expression using the laws of logarithms: $\log_4 \frac{1}{16}$.

A. **–2.** You first apply the rule for the log of a reciprocal and then rewrite the 16 as a power of 4. (The number 16 is also a power of 2, but you want to have the base of the logarithm and the base of the number match so you can take advantage of another property of logs.) Use the rule for the log of a power to bring the 2 in front of the log. Then use the rule for the log of the base to end up multiplying the –2 by 1: $\log_4 \frac{1}{16} = -\log_4 16 = -\log_4 4^2$

$$= -2\log_4 4 = -2(1) = -2.$$

Q. Show how the expression $\log_6 \frac{48}{x}$ is equivalent to $1 + \log_6 8 - \log_6 x$ using the laws of logarithms.

A. First, use the rule for the log of a quotient to write the logarithm as $\log_6 48 - \log_6 x$. Next, factor the number 48 into 6 times 8 to get $\log_6 6 + \log_6 8 - \log_6 x$. Finally, use the rule for the log of the base to change the first term to a 1.

17. Determine which of the following is equivalent to log 45.

  a. log 3 + log 15

  b. log 40 + log 5

  c. 5log 9

  d. 45log 1

18. Determine which of the following is equivalent to $\log \frac{x}{2}$.

  a. $\frac{\log x}{2}$

  b. $\log x + \log 2$

  c. $\log x - 2$

  d. $\log x - \log 2$

19. Determine which of the following is equivalent to $\log_a \frac{1}{3}$.

  a. $\log_a 1 + \log_a 3$

  b. $-\log_a 3$

  c. $-3\log_a 1$

  d. $\frac{1}{\log_a 3}$

20. Determine which of the following is equivalent to ln e.

  a. e

  b. 0

  c. 1

  d. e^e

21. Determine which of the following is equivalent to $\log_6 1$.

  a. 6

  b. 0

  c. 1

  d. –6

22. Determine which of the following is equivalent to log 8.

  a. $\log 4^2$

  b. 3log 2

  c. 2log 3

  d. 4log 2

Presto-Chango: Expanding and Contracting Expressions with Log Functions

The logarithmic expression $\log_2 \frac{4x^3 \sqrt{9x+1}}{(3x^2+7)^5(x-5)}$ is pretty complicated with its numerator, denominator, radical, products, and powers. The beauty of logarithms is that the rules governing them allow you to pull apart and simplify these rather intricately constructed terms. For instance, using the rules of logarithms, you can write the above expression as $2+3\log_2 x+\frac{1}{2}\log_2(9x+1)-5\log_2(3x^2+7)-\log_2(x-5)$. Yes, you have five separate terms now instead of the one, but each of the terms is much simpler — all the logarithms have arguments that are polynomials.

In other mathematics courses, such as calculus, you expand the logarithms in order to have an easier time when computing derivatives. Then, in a later study of math, you take the expanded format and change it back into a single logarithm. You can make all these spectacular changes using the laws of logarithms, which I cover earlier in this chapter, in the section "Checking out the Properties of Logarithms."

Q. Show how to expand the logarithm $\log_2 \dfrac{4x^3\sqrt{9x+1}}{(3x^2+7)^5(x-5)}$ to produce

$2+3\log_2 x+\dfrac{1}{2}\log_2(9x+1)-5\log_2(3x^2+7)-$

$\log_2(x-5)$,

the answer given in the section intro.

A. First, use the rule for the log of a quotient to rewrite the logarithm as $\log_2 4x^3\sqrt{9x+1}-\log_2(3x^2+7)^5(x-5)$. Next, use the rule for the log of a product to split up the factors in the two terms. Be sure to distribute the negative sign over the two terms that come from the second term (the one after the minus sign):

$\log_2 4+\log_2 x^3+\log_2\sqrt{9x+1}-\log_2(3x^2+7)^5$

$-\log_2(x-5)$.

Now change the 4 in the first term to a power of 2 and the radical to a fractional power:

$\log_2 2^2+\log_2 x^3+\log_2(9x+1)^{1/2}$

$-\log_2(3x^2+7)^5-\log_2(x-5)$.

Next, use the rule for the log of a power to bring the 2, 3, $\frac{1}{2}$, and 5 to the front of their respective log terms:

$2\log_2 2+3\log_2 x+\dfrac{1}{2}\log_2(9x+1)-$

$5\log_2(3x^2+7)-\log_2(x-5)$.

Finally, change the $\log_2 2$ to a 1, using the rule for the log of the base. Simplify for the final result:

$2+3\log_2 x+\dfrac{1}{2}\log_2(9x+1)-5\log_2(3x^2+7)-$

$\log_2(x-5)$

Q. Rewrite the terms with logs as a single log expression:

$4\log_7(x-2)+5\log_7(x^2+1)-3\log_7 x-$

$\log_7(x^5-12)-\dfrac{1}{2}\log_7(x^3+2x+1)$

A. $\log_7\dfrac{(x-2)^4(x^2+1)^5}{x^3(x^5-12)\sqrt{x^3+2x+1}}$. First, move

all the coefficients from in front of the logs to respective positions as exponents on the arguments, using the rule for the log of a power:

$\log_7(x-2)^4+\log_7(x^2+1)^5-\log_7 x^3-$

$\log_7(x^5-12)-\log_7(x^3+2x+1)^{1/2}$.

Next, write the first two terms as a product of logs and the last three terms as the product of logs after factoring the negative sign out of each:

$\log_7(x-2)^4(x^2+1)^5-\log_7 x^3(x^5-12)$

$(x^3+2x+1)^{1/2}$.

The last three terms are all negative, so their product ends up in the denominator. Finally, write the expression as a quotient, using the rule for the log of a quotient.

---

23. Expand $\log\dfrac{x^4(x-2)^3}{\sqrt{x+3}}$ using the rules for logarithms.

24. Expand $\log_3\sqrt{\dfrac{x^5(x-2)^3}{(3x+1)^7}}$ using the rules for logarithms.

25. Rewrite $2\log(x-3)+3\log(2x+1)-4\log(x+2)$ $-\frac{1}{2}\log(x-5)$ as a single logarithm using the rules for logarithms.

26. Rewrite $3\ln(x-5)-2\ln(x+4)+3\ln(x+5)$ $-2\ln(x-4)$ as a single logarithm using the rules for logarithms.

Solving Logarithmic Equations

A *logarithmic* equation is one with a log in it. No, I don't mean an oak or birch log, but working with a logarithmic equation can be as easy as falling off that piece of fallen tree! You may think I'm joking, but this *falling off* business is pretty close to the truth.

The best way to solve a logarithmic equation is to get off it — move away from that format. Change a log equation to either an algebraic format or an exponential equation so you can solve it:

✔ **Change the log equation $\log_b x = \log_b y$ to $x = y$.** Then solve the algebraic equation. Notice that only one log term is on each side and that the bases are the same. You set the two arguments equal to one another.

✔ **Change the log equation $\log_b x = y$ to $b^y = x$.** Then solve the exponential equation. Again, notice that the equation has only one log term, that the base of that logarithm becomes the base of the exponential expression, and that the exponential expression (b^y) is set equal to the argument (x).

Whenever you change the format of an equation to make it easier to solve, check for extraneous (false) solutions by inserting your answers in the original equation. Even though not all answers work, this changing-the-format process is still the preferred method for solving most logarithmic equations.

Q. Solve the equation $\log_4(x-5)+\log_4(x+3)$ $=\log_4(5x+3)$.

A. $x = 9$. Before changing the equation to an algebraic format, you have to combine the two terms on the left. For you to use the rule that gets rid of the logs, the equation can have only one log term on each side. Using the rule for the log of a product, you rewrite the equation as $\log_4[(x-5)(x+3)]$ $=\log_4(5x+3)$. Now you can set the two *arguments* of the logarithms equal to one another to get $(x-5)(x+3)=5x+3$. This equation is quadratic. Solve it by first multiplying the two binomials together on the left and then setting the equation equal to 0. The equation $x^2-2x-15=5x+3$

becomes $x^2-7x-18=0$. Factor the quadratic into $(x-9)(x+2)=0$. This quadratic equation has two solutions: $x=9$ or $x=-2$. The first solution, $x=9$, is a valid solution of the original log equation. The equation is true when you replace all the x's in the original equation with 9. You get $\log_4(9-5)+\log_4(9+3)=\log_4(5\cdot 9+3)$; $\log_4(4)+\log_4(12)=\log_4(48)$; $\log_4(4\cdot 12)=\log_4(48)$. The other solution, $x=-2$, doesn't work in the log equation. You get $\log_4(-2-5)+\log_4(-2+3)=\log_4[5(-2)+3]$, which results in negative arguments in the first and last logarithmic terms. The arguments of logarithms must always be positive. The solution $x=-2$ is extraneous.

Q. Solve the equation $\log_4(x^2 + 15) = 3$.

A. $x = \pm 7$. Rewrite the equation as an exponential equation. You get $4^3 = x^2 + 15$. Simplifying, the equation becomes $64 = x^2 + 15$. Subtracting 15 from each side, $x^2 = 49$. Taking the square root of each side, $x = \pm 7$. Both the 7 and –7 work, because the x gets squared, so the argument doesn't become negative.

---

27. Solve for x in $\log_3(x + 3) + \log_3(2x - 1) = \log_3(7x + 1)$.

28. Solve for x in $\log_8(3x - 1) - \log_8(x - 3) = \log_8(x + 2)$.

---

29. Solve for x in $\log_5(2x + 3) + \log_5 x = 1$.

30. Solve for x in $\log (x + 8) + 1 = \log (2 - 3x)$.

---

31. Solve for x in $\ln (3x + 5) + \ln (2x + 3) = \ln (x + 3)$.

32. Solve for x in $\ln (x + 1) - \ln x = 2$.

They Ought to Be in Pictures: Graphing Exponential and Logarithmic Functions

Exponential and logarithmic functions are inverses of one another. Consequently, the graphs of exponential and logarithmic functions are mirror images of one another with respect to the line $y = x$. In Figure 9-1, you see the graph of $y = 2^x$ and $y = \log_2 x$ and how they're symmetric with respect to one another over that diagonal line.

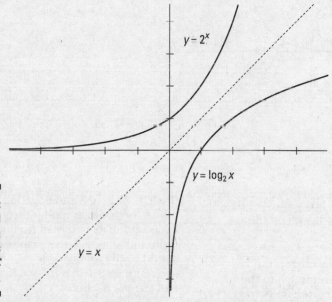

$y = 2^x$

$y = \log_2 x$

$y = x$

Figure 9-1: Exponential and log graphs have soft *C* shapes.

The basic exponential function with a base greater than 1 rises from left to right, getting steeper the farther you move to the right. The basic logarithmic function also rises from left to right, but the curve gets flatter as you move to the right.

You can make the curves fall instead of rise if you use a negative sign. Negating the x in the exponential function makes it fall from left to right, and negating the log in the logarithmic function makes it fall instead of rise. The basic curves can also be made to slide left, right, up, or down by adding or subtracting constants. Here's how you can control the movement:

- ✔ **Up:** Add a constant to the function.
- ✔ **Down:** Subtract a constant from the function.
- ✔ **Left:** Add a constant to the x-term (exponent or argument).
- ✔ **Right:** Subtract a constant from the x-term (exponent or argument).

EXAMPLE

Q. Sketch the graphs of $y = 3^x$, $y = 3^x + 2$, $y = 3^{x-2}$, and $y = 3^{-x}$ all on the same set of axes.

A. Adding 2 to the function $y = 3^x$ raises the basic graph by two units. Subtracting 2 from the x in the exponent part slides the

basic graph two units to the right. Negating the x in the exponent makes the graph fall as the curve moves from left to right. Take a look at Figure 9-2, which shows the curves with these corresponding letters: (A) $y = 3^x$, (B) $y = 3^x + 2$, (C) $y = 3^{x-2}$, and (D) $y = 3^{-x}$.

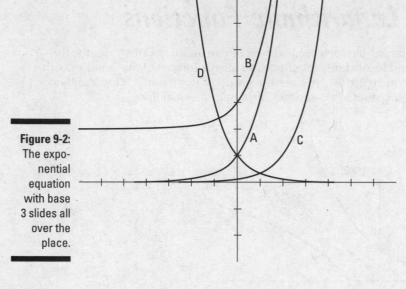

Figure 9-2:
The expo-
nential
equation
with base
3 slides all
over the
place.

EXAMPLE

Q. Sketch the graphs of $y = \ln x$, $y = -3 + \ln x$, $y = \ln (x + 3)$, and $y = 3 - \ln x$ all on the same set of axes.

A. Subtracting 3 from the basic graph lowers it by three units. Adding 3 to the x value moves the graph three units to the left. Adding 3 and negating the $\ln$ raises the basic graph by three units and makes it fall as you read from left to right. In Figure 9-3, the graphs correspond to the following letters: (A) $y = \ln x$, (B) $y = -3 + \ln x$, (C) $y = \ln (x + 3)$, and (D) $y = 3 - \ln x$.

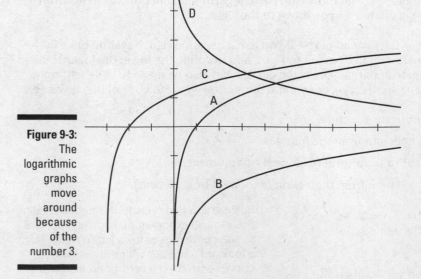

Figure 9-3:
The
logarithmic
graphs
move
around
because
of the
number 3.

33. Sketch the graph of $y = 2^{x-3}$.

34. Sketch the graph of $y = e^{x+2}$.

35. Sketch the graph of $y = \ln(x - 1)$.

36. Sketch the graph of $y = 2 + \ln x$.

37. Sketch the graph of $y = 4^x - 4$.

38. Sketch the graph of $y = 4 + \log(x + 4)$.

Answers to Problems on Exponential and Logarithmic Functions

The following are the answers to the practice problems presented earlier in this chapter.

1 Given the exponential function $f(x) = 2^{x-1}$, determine $f(0)$, $f(1)$, and $f(2)$. The answer is $f(0) = \frac{1}{2}$; $f(1) = 1$; $f(2) = 2$.

Replacing the x's with the input values, you get $f(0) = 2^{0-1} = 2^{-1} = \frac{1}{2}$, $f(1) = 2^{1-1} = 2^0 = 1$, and $f(2) = 2^{2-1} = 2^1 = 2$.

2 Given the exponential function $f(x) = 3^x(1 - 2 \cdot 3^x)$, determine $f(0)$, $f(1)$, and $f(2)$. The answer is $f(0) = -1$; $f(1) = -15$; $f(2) = -153$.

Replacing the x's with the input values, you get $f(0) = 3^0(1 - 2 \cdot 3^0) = 1(1 - 2) = 1(-1) = -1$, $f(1) = 3^1(1 - 2 \cdot 3^1) = 3(1 - 6) = 3(-5) = -15$, and $f(2) = 3^2(1 - 2 \cdot 3^2) = 9(1 - 2 \cdot 9) = 9(1 - 18) = 9(-17) = -153$.

3 Given the exponential function $f(x) = e^x - 2e^{2x}$, determine $f(0)$ and $f(1)$. The answer is $f(0) = -1$; $f(1) = e - 2e^2$.

Replacing the x's with input values, you get $f(0) = e^0 - 2e^{2(0)} = 1 - 2(1) = -1$ and $f(1) = e^1 - 2e^{2(1)} = e - 2e^2$.

4 Given the exponential function $f(x) = (e^x + 2)^2 - (e^x - 2)^2$, determine $f(0)$, $f(1)$, and $f(2)$. The answer is $f(0) = 8$; $f(1) = 8e$; $f(2) = 8e^2$.

Replacing the x's with input values, you have $f(0) = (e^0 + 2)^2 - (e^0 - 2)^2 = (1 + 2)^2 - (1 - 2)^2 = 9 - 1 = 8$. For $f(1)$, you get $(e^1 + 2)^2 - (e^1 - 2)^2$. Now expand each binomial and simplify: $(e^2 + 4e + 4) - (e^2 - 4e + 4) = e^2 + 4e + 4 - e^2 + 4e - 4 = 8e$.

 Instead of squaring the binomials and simplifying, you can factor the expression as the difference of two squares. That's how I handle the last part of the problem.

The difference of two squares factors into the difference and sum of the roots: $a^2 - b^2 = (a - b)(a + b)$.

Here, $f(2) = (e^2 + 2)^2 - (e^2 - 2)^2 = [(e^2 + 2) - (e^2 - 2)][(e^2 + 2) + (e^2 - 2)] = [e^2 + 2 - e^2 + 2][e^2 + 2 + e^2 - 2] = [4][2e^2] = 8e^2$.

5 Given the exponential function $f(x) = 2(e^{x+1})^2$, determine $f(0)$, $f(1)$, and $f(-1)$. The answer is $f(0) = 2e^2$; $f(1) = 2e^4$; $f(-1) = 2$.

Replacing the x's with input values, you get $f(0) = 2(e^{0+1})^2 = 2(e^1)^2 = 2e^2$, $f(1) = 2(e^{1+1})^2 = 2(e^2)^2 = 2e^4$, and $f(-1) = 2(e^{-1+1})^2 = 2(e^0)^2 = 2(1)^2 = 2$.

6 Given the exponential function $f(x) = x^e - e^x$, determine $f(0)$, $f(1)$, and $f(2)$. The answer is $f(0) = -1$; $f(1) = 1 - e$; $f(2) = 2^e - e^2$.

Replacing the x's with input values, you get $f(0) = 0^e - e^0 = 0 - 1 = -1$, $f(1) = 1^e - e^1 = 1 - e$, and $f(2) = 2^e - e^2$.

7 Solve for x in $2^x = 4^{x+1}$. The answer is $x = -2$.

First, rewrite the 4 on the right as 2^2 and simplify the power to get $2^x = 2^{2(x+1)} = 2^{2x+2}$. Now set the two exponents equal to one another: $x = 2x + 2$. Solving for x, $-x = 2$; $x = -2$.

8 Solve for x in $9^{x-5} = \frac{1}{27^x}$. The answer is $x = 2$.

Both the 9 and 27 are powers of 3, so rewrite the terms and simplify the exponents:

$$\left(3^2\right)^{x-5} = \frac{1}{\left(3^3\right)^x}$$

$$3^{2x-10} = \frac{1}{3^{3x}} = 3^{-3x}$$

Setting the exponents equal to one another and solving, $2x - 10 = -3x$; $5x = 10$; $x = 2$.

9 Solve for x in $e^{4x-2} = e^{x+7}$. The answer is $x = 3$.

The bases are the same, so set the exponents equal to one another: $4x - 2 = x + 7$; $3x = 9$; $x = 3$.

10 Solve for x in $2^{2x} - 5 \cdot 2^x + 4 = 0$. The answer is $x = 0, 2$.

This problem is a quadratic-like trinomial that you can factor into the product of two binomials. Let $y = 2^x$. For a pattern, think of the trinomial $y^2 - 5y + 4$, which factors into $(y-4)(y-1)$. Factoring the trinomial with the exponential terms, you get $(2^x - 1)(2^x - 4) = 0$. Setting the first factor equal to 0, you get $2^x - 1 = 0$; $2^x = 1$; $x = 0$. With the second factor, you calculate $2^x - 4 = 0$; $2^x = 4$. The 4 is a power of 2, so write the equation as $2^x = 2^2$; thus, $x = 2$. Both solutions work.

11 Solve for x in $e^x - 1 = 0$. The answer is $x = 0$.

Rewrite the equation as $e^x = 1$. Think of 1 as being equal to e^0; thus, $e^x = e^0$; $x = 0$.

12 Solve for x in $xe^x + 5e^x = 0$. The answer is $x = -5$.

The binomial factors — the common factor is e^x. So $e^x(x + 5) = 0$. Using the multiplication property of zero, you set each of the factors equal to 0 and solve for x. But $e^x = 0$ has no solution. No power of e is equal to 0, so this factor doesn't yield a solution. However, $x + 5 = 0$ when $x = -5$.

13 Determine the total amount of money accumulated after 20 years if you deposit \$1,000,000 in an account that earns 5.25% interest, compounded monthly. The answer is **\$2,851,114.02.**

In 20 years, the money almost triples. Using the formula for compound interest, you write

$$A = 1,000,000\left(1 + \frac{.0525}{12}\right)^{12(20)}$$

$$= 1,000,000(1.004375)^{240}$$

$$\approx 2,851,114.02$$

14 Determine the total amount of money accumulated after 2 years and 6 months if you deposit \$25,000 in an account that earns 5% interest, compounded daily (use 365 days). The answer is **\$28,328.47.**

Six months is half a year, so write 2 years, 6 months as 2.5 years. Use the compound interest formula:

$$A = 25,000\left(1 + \frac{.05}{365}\right)^{365(2.5)}$$

$$\approx 25,000(1.000136986)^{912.5}$$

$$\approx 28,328.47$$

15 Determine the total amount of money accumulated after 20 years if you deposit \$1,000,000 in an account that earns 5.25% interest, compounded continuously. The answer is **\$2,857,651.12.**

Using the formula for compounding continuously, $A = 1,000,000e^{0.0525(20)} = 1,000,000e^{1.05} \approx 2,857,651.12$. You get a difference of over \$6,000 between compounding monthly and compounding continuously (refer to problem 13 for the calculation that involves monthly compounding).

16 Determine the total amount of money accumulated after 2 years and 6 months if you deposit $25,000 in an account that earns 5% interest, compounded continuously. The answer is **$28,328.71.**

Using the formula for compounding continuously, $A = 25,000e^{0.05(2.5)} = 1,000,000e^{0.125} \approx 28,328.71$. When you compare this answer to the result in problem 14, you don't find much difference between compounding continuously and compounding daily.

17 Determine which of the following is equivalent to log 45. The answer is **a. log 3 + log 15.**

Because 45 is the product of 3 · 15, you can write log 45 = log (3 · 15) and then apply the rule for the log of a product.

18 Determine which of the following is equivalent to $\log \frac{x}{2}$. The answer is **d. log x – log 2.**

The expression involves a quotient, so apply the rule for the log of a quotient.

19 Determine which of the following is equivalent to $\log_a \frac{1}{3}$. The answer is **b. –log$a$3.**

Two different approaches to this problem are available. You can apply the rule for the log of a quotient and write the expression as $\log_a 1 - \log_a 3$ and then replace $\log_a 1$ with 0, using the rule for the log of 1. Simplifying $0 - \log_a 3$, you get $-\log_a 3$. The easier approach is to just know the rule for the log of a reciprocal and apply it. The easier or simpler way isn't always the most obvious, so either way is really acceptable.

20 Determine which of the following is equivalent to ln e. The answer is **c. 1.**

Just apply the rule for the log of the base. Technically, you don't have to show the base when the base is e. You're working with the natural logarithm. Rewriting as an exponential equation, you have $e^x = e$ which is $e^x = e^1$, so $x = 1$.

21 Determine which of the following is equivalent to $\log_6 1$. The answer is **b. 0.**

This problem follows the rule for the log of 1.

22 Determine which of the following is equivalent to log 8. The answer is **b. 3log 2.**

First, change the 8 to a power of 2, and then apply the rule for the log of a power. You get $\log 8 = \log 2^3 = 3\log 2$.

23 Expand $\log \frac{x^4(x-2)^3}{\sqrt{x+3}}$ using the rules for logarithms. The answer is $4\log x + 3\log(x-2) - \frac{1}{2}(x+3)$.

First, rewrite the log using the rule for the log of a quotient to get $\log x^4(x-2)^3 - \log \sqrt{x+3}$. Now use the rule for the log of a product on the first term, and then change the radical to a fractional power: $\log x^4 + \log(x-2)^3 - \log(x+3)^{1/2}$. Use the rule for the log of a power to rewrite each term without an exponent.

24 Expand $\log_3 \sqrt{\frac{x^5(x-2)^3}{(3x+1)^7}}$ using the rules for logarithms. The answer is

$\frac{5}{2}\log_3 x + \frac{3}{2}\log_3(x-2) - \frac{7}{2}\log_3(3x+1)$.

Change the radical to an exponent, and then apply the rule for the log of a power, moving the $\frac{1}{2}$ in front of the ln. Then apply the rule for a quotient; be sure to use a grouping symbol so that each term is multiplied by the $\frac{1}{2}$ factor in the front: $\frac{1}{2}\log_3 \frac{x^5(x-2)^3}{(3x+1)^7} = \frac{1}{2}\left[\log_3 x^5(x-2)^3 - \log(3x+1)^7\right]$.

Now apply the rule for a product, and then move the powers in front of the respective *logs*.

Lastly, distribute the $\frac{1}{2}$ through all the terms:

$$= \frac{1}{2} \left[\log_3 x^5 + \log_3 (x-2)^3 - \log (3x+1)^7 \right]$$

$$= \frac{1}{2} \left[5\log_3 x + 3\log_3 (x-2) - 7\log(3x+1) \right]$$

$$= \frac{5}{2} \log_3 x + \frac{3}{2} \log_3 (x-2) - \frac{7}{2} \log(3x+1)$$

25 Rewrite $2\log(x-3) + 3\log(2x+1) - 4\log(x+2) - \frac{1}{2}\log(x-5)$ as a single logarithm using the rules for logarithms. The answer is $\log \dfrac{(x-3)^2(2x+1)^3}{(x+2)^4\sqrt{x-5}}$.

First, rewrite each term with its multiplier changed to an exponent. Then group the first two terms and rewrite them using the rule for the log of a product. Factor the negative sign out of the second two terms and apply the rule for a product again. Lastly, use the rule for a quotient and change the binomial with the $\frac{1}{2}$ exponent into a binomial under a radical:

$$\log(x-3)^2 + \log(2x+1)^3 - \log(x+2)^4 - \log(x-5)^{1/2}$$

$$= \log(x-3)^2(2x+1)^3 - \left[\log(x+2)^4 + \log(x-5)^{1/2} \right]$$

$$= \log(x-3)^2(2x+1)^3 - \left[\log(x+2)^4(x-5)^{1/2} \right]$$

$$= \log \frac{(x-3)^2(2x+1)^3}{(x+2)^4(x-5)^{1/2}} = \log \frac{(x-3)^2(2x+1)^3}{(x+2)^4\sqrt{x-5}}$$

26 Rewrite $3\ln(x-5) - 2\ln(x+4) + 3\ln(x+5) - 2\ln(x-4)$ as a single logarithm using the rules for logarithms. The answer is $\log_8 \dfrac{(x^2-25)^3}{(x^2-16)^2}$.

Rewrite each term by moving the multiplier in front to its place as an exponent. Next, rearrange the terms, grouping the positive and negative, and factoring the negative sign out of the negative terms. Apply the rule for the log of a product for the two groupings, and then apply the rule for the log of a quotient. Notice that the binomials in the numerator have the same power, as do the binomials in the denominator. The binomials are the sum and difference of the same two numbers, so you can write them as the difference of squares:

$$\log_8(x-5)^3 - \log_8(x+4)^2 + \log_8(x+5)^3 - \log_8(x-4)^2$$

$$= \left[\log_8(x-5)^3 + \log_8(x+5)^3 \right] - \left[\log_8(x+4)^2 + \log_8(x-4)^2 \right]$$

$$= \left[\log_8(x-5)^3(x+5)^3 \right] - \left[\log_8(x+4)^2(x-4)^2 \right]$$

$$= \log_8 \frac{(x-5)^3(x+5)^3}{(x+4)^2(x-4)^2} = \log_8 \frac{(x^2-25)^3}{(x^2-16)^2}$$

27 Solve for x in $\log_3(x+3) + \log_3(2x-1) = \log_3(7x+1)$. The answer is **$x = 2$**.

First, use the rule for the log of a product to rewrite the left side of the equation, giving you $\log_3[(x+3)(2x-1)] = \log_3(7x+1)$. This equation is now in the form $\log x = \log y$, so you can set the two arguments equal to one another and solve the algebraic equation: $(x+3)(2x-1) = 7x+1$. Multiply on the left. Then set the equation equal to 0 and factor. The equation $2x^2 + 5x - 3 = 7x + 1$ becomes $2x^2 - 2x - 4 = 0$. The polynomial factors into $2(x^2 - x - 2) = 2(x-2)(x+1) = 0$. The two solutions of the quadratic equation are $x = 2$ and $x = -1$. The answer -1 doesn't work, because it gives you logs of negative numbers in the original equation. Thus, the solution $x = -1$ is extraneous.

28 Solve for x in $\log_8(3x - 1) - \log_8(x - 3) = \log_8(x + 2)$. The answer is **$x = 5$.**

Apply the rule for the log of a quotient to rewrite the equation as $\log_8 \dfrac{3x - 1}{x - 3} = \log_8(x + 2)$. Then set the two arguments equal to one another and multiply each side by $x - 3$:

$$\frac{3x - 1}{x - 3} = x + 2$$

$$(x - 3)\frac{3x - 1}{x - 3} = (x + 2)(x - 3)$$

$$3x - 1 = x^2 - x - 6$$

Setting the equation equal to 0, you get $0 = x^2 - 4x - 5 = (x - 5)(x + 1)$. The two solutions of the quadratic are $x = 5$ and $x = -1$, but only the $x = 5$ works — the other is extraneous.

29 Solve for x in $\log_5(2x + 3) + \log_5 x = 1$. The answer is **$x = 1$.**

Use the rule for the log of a product to write the equation as $\log_5[(2x + 3)x] = 1$. Now change this format to $5^1 = (2x + 3)x$. Distribute the x on the right, and then set the equation equal to 0, giving you $5 = 2x^2 + 3x; 0 = 2x^2 + 3x - 5$. The quadratic factors: $0 = (2x + 5)(x - 1)$. The two solutions are $x = -\dfrac{5}{2}$ and $x = 1$, but only the $x = 1$ works — the other is extraneous.

30 Solve for x in $\log(x + 8) + 1 = \log(2 - 3x)$. The answer is **$x = -6$.**

First subtract $\log(x + 8)$ from each side, and then apply the rule for the log of a quotient. Rewrite the equation in exponential format; remember that $\log x$ is a common logarithm — the base is 10:

$$1 = \log(2 - 3x) - \log(x + 8)$$

$$1 = \log \frac{2 - 3x}{x + 8}$$

$$10^1 = \frac{2 - 3x}{x + 8}$$

Now multiply each side of the equation by $x + 8$ and solve for x:

$$10(x + 8) = 2 - 3x$$

$$10x + 80 = 2 - 3x$$

$$13x = -78$$

$$x = -6$$

31 Solve for x in $\ln(3x + 5) + \ln(2x + 3) = \ln(x + 3)$. The answer is **$x = -1$.**

Rewrite the left side using the rule for the log of a product: $\ln[(3x + 5)(2x + 3)] = \ln(x + 3)$. Then use the rule where two logs are equal to create the algebraic equation: $(3x + 5)(2x + 3) = x + 3$. Multiply on the left; then set the equation equal to 0 by adding $-x$ and -3 to each side of the equation. Then factor: $6x^2 + 19x + 15 = x + 3; 6x^2 + 18x + 12 = 0; 6(x^2 + 3x + 2) = 6(x + 1)(x + 2) = 0$. The two solutions are $x = -1$ and $x = -2$. The solution $x = -1$ works, but the solution $x = -2$ is extraneous.

32 Solve for x in $\ln (x + 1) - \ln x = 2$. The answer is $x = \dfrac{1}{e^2 - 1}$.

Rewrite the left side using the rule for the log of a quotient. Then write the exponential form of the equation.

 The base of the logarithm designated by *ln* is base e.

$$\ln \frac{x+1}{x} = 2$$

$$e^2 = \frac{x+1}{x}$$

Multiply each side by x, and then solve for x:

$$e^2 x = x + 1$$

$$e^2 x - x - 1$$

$$x(e^2 - 1) = 1$$

$$x = \frac{1}{e^2 - 1}$$

33 Sketch the graph of $y = 2^{x-3}$.

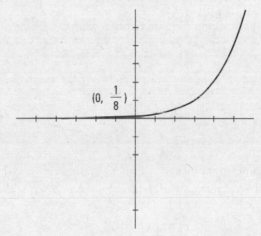

$(0, \frac{1}{8})$

The y-intercept is $\left(0, \dfrac{1}{8}\right)$. The graph of this function is the basic exponential graph of $y = 2^x$ moved three units to the right.

34 Sketch the graph of $y = e^{x+2}$.

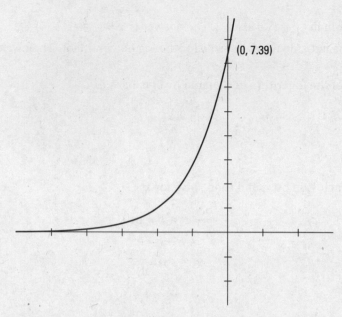

The *y*-intercept is $(0, e^2)$ which is about $(0, 7.39)$. The graph of this function is the basic exponential graph of $y = e^x$ moved two units to the left.

35 Sketch the graph of $y = \ln(x - 1)$.

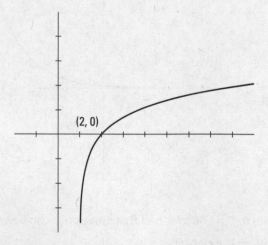

The *x*-intercept is $(2, 0)$. The graph of this function is the basic logarithmic graph of $y = \ln x$ moved one unit to the right.

36 Sketch the graph of $y = 2 + \ln x$.

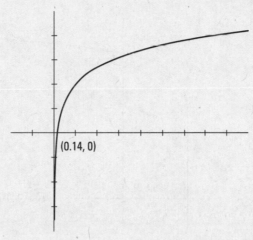

The *x*-intercept is $\left(\frac{1}{e^2}, 0\right)$ which is about (0.14,0). The graph of this function is the basic logarithmic graph of $y = \ln x$ moved two units up.

37 Sketch the graph of $y = 4^x - 4$

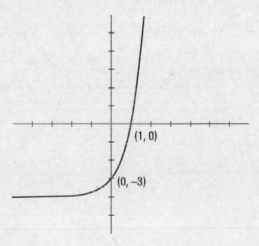

The two intercepts are (0,–3) and (1,0). The graph of this function is the basic exponential graph of $y = 4^x$ lowered by four units.

38 Sketch the graph of $y = 4 + \log (x + 4)$.

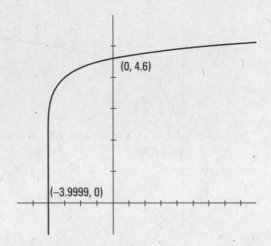

The two intercepts are $(0, 4 + \log 4)$ and $(-3.9999, 0)$, which are approximately $(0, 4.6)$ and $(-4, 0)$. You get the y-intercept by setting x equal to 0, giving you $y = 4 + \log 4$. You solve for the x-intercept by letting $y = 0$ and solving $0 = 4 + \log(x + 4)$, which becomes $\log(x + 4) = -4$; $10^{-4} = x + 4$; $x = -4 + 10^{-4}$. The graph of this function is the basic logarithmic graph of $y = \log x$ moved up four units and left four units.

Part III
Conics and Systems of Equations

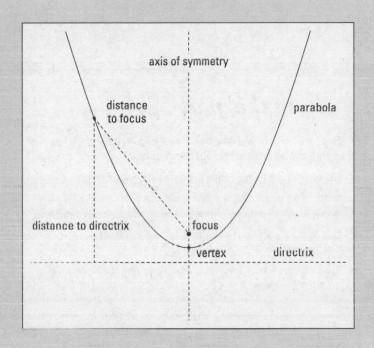

In this part . . .

- Become familiar with the conic sections — what they have in common and what makes them different.

- Sketch conics using centers, intercepts, foci, and other properties.

- Solve linear equations algebraically or with Cramer's rule.

- Use systems of linear equations to perform fraction decomposition.

- Take on systems of nonlinear equations and their possible multiple solutions.

Chapter 10

Any Way You Slice It: Conic Sections

● ●

In This Chapter

▶ Writing equations of parabolas, circles, ellipses, and hyperbolas in standard form

▶ Identifying centers, foci, and axes of conic sections

▶ Sketching parabolas, circles, ellipses, and hyperbolas

● ●

Chapter 6 contains information on quadratic functions, which you graph as U-shaped curves called *parabolas*. Here, you see that not all parabolas represent functions. When a parabola opens to the left or right, you get more than one *y*-value for many *x*-values, and that just doesn't work with functions. This chapter introduces parabolas and some other conic sections that don't exactly fit the function mold. All parabolas are conic sections, but not all parabolas are functions — it depends on how they lie.

A *conic section,* or *conic,* is represented by a slice of a cone (or really, two cones stacked on top of each other, like an hourglass). How you slice the cone — at what angle and how big a chunk you take — determines what kind of conic section you produce. The four types of conic sections are the parabola, circle, ellipse, and hyperbola. Here's how they compare:

✔ **Parabola:** You get this *U*-shaped curve when you take a slice through a cone down to the base.

✔ **Circle:** A circle appears when you cut a cone straight across, parallel to the base.

✔ **Ellipse:** If you slice through a cone at a slant, you get an ellipse, or oval.

✔ **Hyperbola:** Slicing straight down through both cones gives you a hyperbola. It looks like two *U*-shaped curves facing away from each other.

In this chapter, you use the standard forms of equations of conics to recognize which type of conic section an equation represents. You also find the defining characteristics of the conic sections and use these traits to help you graph the conics.

Putting Equations of Parabolas in Standard Form

Rules representing parabolas come in two standard forms to separate the *functions* opening upward or downward from *relations* that open sideways. The standard forms tell you what the parabola looks like — its general width or narrowness, in which direction it opens, and where the vertex (turning point) of the graph is.

Here are the two standard forms for the equations of parabolas (and what they tell you):

Equation	*Direction the Parabola Opens*	*Location of the Vertex*
$(y - k)^2 = 4a(x - h)$	Right if a is positive, left if a is negative	(h, k)
$(x - h)^2 = 4a(y - k)$	Up if a is positive, down if a is negative	(h, k)

The multiplier $4a$ is a constant that tells you how steep or wide the parabola is. You can use that information in the next section, "Shaping Up: Determining the Focus and Directrix of a Parabola."

Q. Write the equation of the parabola $x^2 - 16x - 4y + 52 = 0$ in standard form to determine its vertex and in which direction it opens.

A. $(x - 8)^2 = 4(1)(y + 3)$; **vertex: (8, –3); opens upward.** First rewrite the equation to isolate the x-terms: Leave the two terms with x's on the left, and get the other two terms on the right by adding $4y$ and -52 to both sides. You get $x^2 - 16x = 4y - 52$. Next, complete the square (go to Chapter 2 if you need to know more about this technique) on the left side of the equation. Be sure to add the 64 to both sides of the equation to keep it balanced; $x^2 - 16x + 64 = 4y - 52 + 64$ becomes $(x - 8)^2 = 4y + 12$. Now factor out the 4 from the terms on the right to get $(x - 8)^2 = 4(y + 3)$. The vertex of the parabola is (8, –3). The parabola opens upward, because the x term is squared and the multiplier on the right is positive. The $4a$ part of the standard form is actually 4(1), if you want to show that the a value is 1. (You need to know information about the a in the next section.)

Q. Write the equation of the parabola $2y^2 + 28y + x + 97 = 0$ in standard form to determine its vertex and in which direction it opens.

A. $(y + 7)^2 = (x - 1)$; **vertex: (1, –7); opens left.** First, rewrite the equation, leaving the two terms with y's on the left and moving the others to the right. You get $2y^2 + 28y = -x - 97$. Factor the 2 out of the terms on the left before completing the square; $2(y^2 + 14y) = -x - 97$ becomes $2(y^2 + 14y + 49) = -x - 97 + 98$. Notice that you have to add 98 to the right, because the 49 that you added to complete the square is multiplied by 2. Simplifying and factoring, you have $2(y + 7)^2 = -x + 1$. Now factor –1 from each term on the right, and then divide both sides by 2: $2(y + 7)^2 = -1(x - 1)$, $(y + 7)^2 = -\frac{1}{2}(x - 1)$. The vertex of the parabola is at (1, –7), and the parabola opens to the left. The coefficient on the right, $-\frac{1}{2}$, is written $4\left(-\frac{1}{8}\right)$ when you want to put it in the $4a$ form.

1. Write the equation of the parabola $x^2 - 6x - y + 4 = 0$ in standard form to determine its vertex and in which direction it opens.

2. Write the equation of the parabola $2x^2 + 8x + y + 3 = 0$ in standard form to determine its vertex and in which direction it opens.

3. Write the equation of the parabola $3y^2 - 12y - x + 9 = 0$ in standard form to determine its vertex and in which direction it opens.

4. Write the equation of the parabola $y^2 - x + 4 = 0$ in standard form to determine its vertex and in which direction it opens.

5. Write the equation of the parabola $y^2 + y + x = 0$ in standard form to determine its vertex and in which direction it opens.

6. Write the equation of the parabola $4x^2 - x - y + 1 = 0$ in standard form to determine its vertex and in which direction it opens.

Shaping Up: Determining the Focus and Directrix of a Parabola

A *parabola* is technically defined as all the points that are the same distance from a fixed point called the *focus* and a fixed line called the *directrix*. Both elements are related to the parabola's axis of symmetry. The directrix is always perpendicular to the parabola's axis of symmetry, and the vertex of the parabola lies on the axis of symmetry and is exactly halfway between the focus and the directrix. Figure 10-1 shows you a parabola with all these features identified.

You can identify the focus and directrix (as well as the vertex and axis of symmetry) of a parabola from the standard form of the parabola's equation:

Standard Form	Focus	Directrix	Vertex	Axis of Symmetry
$(y - k)^2 = 4a(x - h)$	$(h + a, k)$	$x = h - a$	(h, k)	$y = k$
$(x - h)^2 = 4a(y - k)$	$(h, k + a)$	$y = k - a$	(h, k)	$x = h$

So, what can foci do for you in the real world? Well, if you're trying to concentrate the sun's rays with a parabolic mirror so you can set a fleet of Roman ships on fire, you have to know where your target should be, right? (A pretty cool idea in theory, if not in practice. We don't know whether Archimedes pulled off the attack back in 212 B.C., but the concept does work for harnessing solar energy in power plants!)

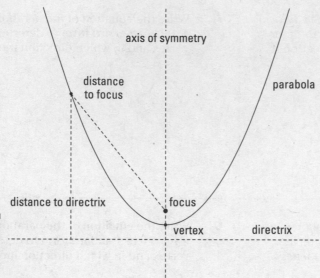

axis of symmetry

distance
to focus

parabola

distance to directrix

focus

Figure 10-1:
The focus
is inside the
parabola.

vertex   directrix

EXAMPLE

Q. Find the vertex, focus, directrix, and axis of symmetry for the parabola $x^2 = -8y$.

A. **Vertex: (0, 0); focus: (–2, 0); directrix: $y = 2$; axis of symmetry: $x = 0$.** The standard form for this parabola, indicating the values of h, k, and a, is written $(x - 0)^2 = 4(-2)(y - 0)$. The value of a is –2. The vertex of the parabola is at (0, 0), and the focus is at $(0 - 2, 0) = (-2, 0)$. The directrix is $y = 0 - (-2)$ or $y = 2$, and the axis of symmetry is $x = 0$.

Q. Find the vertex, focus, directrix, and axis of symmetry for the parabola $(y - 2)^2 = 24(x + 3)$.

A. **Vertex: (–3, 2); focus: (3, 2); directrix: $x = -9$; axis of symmetry: $y = -2$.** The standard form for this parabola, indicating the values of h, k, and a, is written $(y - 2)^2 = 4(6)(x + 3)$. (***Note:*** In other problems, you may have to complete the square to put the equation in standard form. See the preceding section and Chapter 2 for more information.) The value of a is 6. The vertex of the parabola is at (–3, 2), and the focus is at $(-3 + 6, 2) = (3, 2)$. The directrix is $x = -3 - 6$ or $x = -9$, and the axis of symmetry is $y = -2$.

7. Find the vertex, focus, directrix, and axis of symmetry for the parabola $x^2 = 16y$.

8. Find the vertex, focus, directrix, and axis of symmetry for the parabola $y^2 = -100x$.

9. Find the vertex, focus, directrix, and axis of symmetry for the parabola $8x^2 - 32x + 38 - y = 0$.

10. Find the vertex, focus, directrix, and axis of symmetry for the parabola $y^2 - 2y + 4x + 13 = 0$.

Back to the Drawing Board:
Sketching Parabolas

A parabola is a smooth, *U*-shaped curve, and you can sketch a parabola using all the valuable information gleaned from its equation in the standard form. The vertex, focus, directrix, and axis of symmetry are items you use in the sketch. You can also solve for intercepts and other random points if you need more information.

The value of *a* in the standard form of the parabola determines how wide or steep the parabola is and what direction it faces:

✔ When *a* is positive, the parabola opens upward or to the right.

✔ When *a* is negative, the parabola opens downward or to the left.

✔ The bigger the absolute value of *a*, the wider or flatter the parabola. As the absolute value of *a* gets larger, the focus and directrix get farther apart.

✔ When *a* is a number between –1 and +1 (when its absolute value is a proper fraction but not 0), the parabola steepens — the focus and directrix get pretty close together, and the parabola gets steeper or more compact.

Refer to the preceding section for info on finding the vertex, focus, directrix, and axis of symmetry. (I cover finding intercepts of quadratic functions in Chapter 2.)

Q. Sketch the graph of $(y - 3)^2 = 12(x + 1)$. Label the vertex, focus, directrix, and axis of symmetry.

A. As Figure 10-2 shows, the vertex is (–1, 3); the value of *a* is 3; the focus is (–1 + 3, 3) = (2, 3); the directrix is $x = -1 - 3 = -4$; and the axis of symmetry is $y = 3$. The parabola opens to the right, because *a* is positive.

Q. Sketch the graph of $(x - 5)^2 = -(y - 3)$. Label the vertex, focus, directrix, and axis of symmetry.

A. Check out Figure 10-3. The vertex is (5, 3); the value of *a* is $-\frac{1}{4}$; the focus is $\left(5, 3 - \frac{1}{4}\right) = \left(5, 2\frac{3}{4}\right)$; the directrix is $y = 3 - \left(-\frac{1}{4}\right) = 3\frac{1}{4}$; and the axis of symmetry is $x = 5$. The parabola opens downward, because *a* is negative.

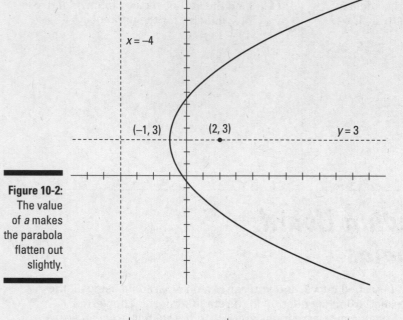

Figure 10-2:
The value of *a* makes the parabola flatten out slightly.

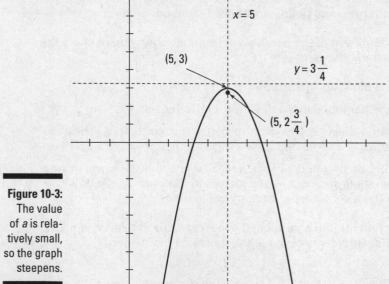

Figure 10-3:
The value of *a* is relatively small, so the graph steepens.

11. Sketch the graph of $(x - 4)^2 = 12(y + 1)$. Label the vertex, focus, directrix, and axis of symmetry.

12. Sketch the graph of $(y - 1)^2 = 8(x - 5)$. Label the vertex, focus, directrix, and axis of symmetry.

13. Sketch the graph of $4y^2 + 24y + 16x + 33 = 0$. Label the vertex, focus, directrix, and axis of symmetry.

14. Sketch the graph of $2x^2 + 8x + 11 = y$. Label the vertex, focus, directrix, and axis of symmetry.

Writing the Equations of Circles and Ellipses in Standard Form

A *circle* is all the points that are a constant distance, or *radius,* from a fixed point called its *center.*

An ellipse is an oval that's taller than wide or wider than tall, and it's based around two fixed points called *foci* (the plural of *focus*). An *ellipse* consists of all the points where the sum of the distances from a point on the ellipse to the *foci* is constant.

The foci lie along the ellipse's *major axis,* which is the longest line you can draw inside the ellipse. A *minor axis* is perpendicular to the major axis and intersects the major axis at the *center* of the ellipse. Figure 10-4a shows a circle with its center and radius (r) identified. Figure 10-4b has an ellipse with its axes, foci, and segments drawn from one point to the two foci.

Here are the standard forms for circles and ellipses:

- **Circle:** $(x - h)^2 + (y - k)^2 = r^2$, where

  - (h, k) is the center

  - r is the radius

- **Ellipse:** $\dfrac{(x - h)^2}{a^2} + \dfrac{(y - k)^2}{b^2} = 1$, where

  - (h, k) is the center

  - $2a$ is the length of one axis and $2b$ is the length of the other (the major axis corresponds to whichever is larger, a or b)

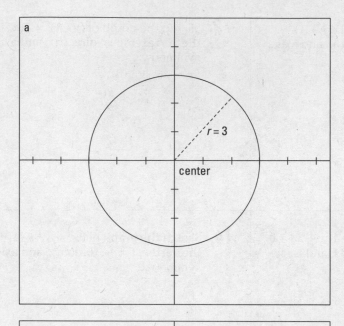

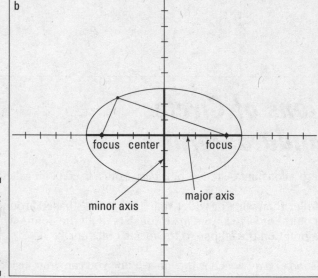

Figure 10-4:
A circle and
an ellipse, a
"squished"
circle.

Q. Write the equation of the circle $x^2 + y^2 - 12x + 24y = 45$ in standard form. Identify the center and radius of the circle.

A. $(x - 6)^2 + (y + 12)^2 = 225$; **center: (6, –12); radius: 15 units.** You have to complete the square twice (see Chapter 2). First, group the x and y terms and determine what has to be added to each grouping to create a perfect square trinomial. You get $(x^2 - 12x + 36) + (y^2 + 24y + 144) = 45 + 36 + 144$. Be sure to add the same amount to each side of the equation. Factoring on the left and simplifying on the right, you get $(x - 6)^2 + (y + 12)^2 = 225$. The center of the circle is (6, –12), and the radius is 15 units.

Q. Write the equation of the ellipse $25x^2 + 150x + 9y^2 + 18y + 9 = 0$ in standard form. Identify the center, the length of the major axis, and the length of the minor axis.

A. $\dfrac{(x+3)^2}{9} + \dfrac{(y+1)^2}{25} = 1$; center: $(-3, -1)$; **major axis: 10 units; minor axis: 6 units.** You have to complete the square twice to create the standard form. Factor the 25 out of the first two terms and 9 out of the second two; then subtract 9 from each side: $25(x^2 + 6x) + 9(y^2 + 2y) = -9$. To create a perfect square trinomial in the first set of parentheses, add a 9 after the $6x$ and add 225 to the right side of the equation (the 25 multiplies the 9, also). To create a perfect square trinomial in the second set of parentheses, add 1 after the $2y$ and 9 to the right side. Factoring and simplifying, you have $25(x + 3)^2 + 9(y + 1)^2 = 225$. Now divide each term by 225 and simplify to get the standard form. The center of the ellipse is $(-3, -1)$. The major axis is 10 units long. You get this answer by taking the square root of the larger of the two denominators, the 25, and doubling it. The minor axis is 6 units long — the square root of 9 is 3, and twice 3 is 6. This ellipse is taller than it is wide, because the larger denominator is under the y-values.

---

15. Write the equation of the circle $x^2 + y^2 - 6x + 8y = 11$ in standard form. Identify the center and radius of the circle.

16. Write the equation of the circle $x^2 + y^2 + 2x - 10y + 17 = 0$ in standard form. Identify the center and radius of the circle.

17. Write the equation of the ellipse $9x^2 + 4y^2 - 90x - 16y + 205 = 0$ in standard form. Identify the center and the lengths of the two axes.

18. Write the equation of the ellipse $x^2 + 16y^2 + 4x - 32y + 19 = 0$ in standard form. Identify the center and the lengths of the two axes.

Determining Foci and Vertices of Ellipses

The *foci* of an ellipse are two points that lie on the major axis (the longer of the two axes). The foci are sort of anchors or determiners for the ellipse. You measure the distances from a particular point on the ellipse to each focus and add up the distances. That sum is the same for any point on the ellipse.

The closer to the center the foci are, the closer the ellipse is to a circle. You can determine the distance (c) from the center to each focus with the formula $c^2 = a^2 - b^2$ or $c^2 = b^2 - a^2$, depending on whether a or b is larger — you want the difference to be positive (in other words, you want the absolute value of $a^2 - b^2$). The a^2 and b^2 are the values in the denominator of the standard form for the ellipse, $\frac{(x-h)^2}{a^2} + \frac{(y-k)^2}{b^2} = 1$. They represent half the length of the axes. The value of c comes out to be $\pm c$ when you take the square root, and those numbers give you the distances of the foci on either side of the center of the ellipse.

The *vertices* of an ellipse are the two endpoints of the major axis. To find these points, you add and subtract the value of a or b — whichever one corresponds to the major axis — from the coordinates of the center (h, k) of the ellipse. Do the addition and subtraction to the x part of the coordinate if the major axis is horizontal, and do the computations to the y part of the coordinate if the major axis is vertical.

Q. Find the foci and vertices of the ellipse $\frac{(x+3)^2}{49} + \frac{(y-5)^2}{625} = 1$.

A. **Foci: (–3, 29), (–3, –19); vertices: (–3, 30), (–3, –20).** The foci lie on the major axis, and the major axis is the vertical axis in this case — the larger denominator is under the y-values. Using $c^2 = b^2 - a^2$, you get $c^2 = 625 - 49 = 576$. So $c = \pm 24$. To find the foci, add and subtract 24 from

the y-value of the center of the ellipse, (–3, 5). The foci are at (–3, 29) and (–3, –19). The vertices are on the same axis and are 25 units (b units) above and below the center. The vertices are at (–3, 30) and (–3, –20). With the foci so close to the vertices, you have a very tall and narrow ellipse. (Check out the graph of this ellipse in the second example of the next section.)

---

19. Find the foci and vertices of the ellipse $\frac{(x+6)^2}{16} + \frac{(y-2)^2}{25} = 1$.

20. Find the foci and vertices of the ellipse $\frac{(x-1)^2}{625} + \frac{(y-7)^2}{576} = 1$.

21. Find the foci and vertices of the ellipse $\frac{x^2}{169} + \frac{y^2}{144} = 1$.

22. Find the foci and vertices of the ellipse $100x^2 + 36y^2 = 3,600$.

Rounding Out Your Sketches: Circles and Ellipses

Sketching a circle is really rather simple — if you have the standard form of its equation (see the earlier section "Writing the Equations of Circles and Ellipses in Standard Form"). You find the center of the circle and determine its radius. Then you mark the center on your graph paper; find points that are the distance of the radius above, below, to the right, and to the left of the center; and draw in the rest of the circle.

Sketching an ellipse isn't all that much harder than drawing a circle. You find the center, the vertices (the endpoints of the major axis), and then the two endpoints of the minor axis. The ellipse is sort of like a circle, except the horizontal and vertical distances are different from one another. Again, you need the equation of the ellipse in its standard form.

Q. Sketch the circle $9(x - 1)^2 + 9(y + 4)^2 = 49$, indicating the center and radius on the graph.

A. First, divide each term by 9 to put the equation of the circle in its standard form. The center of the circle is at $(1, -4)$, and the radius is $\frac{7}{3}$ — just a little over 2 units (see Figure 10-5). Mark the center and then estimate the distance about 2 units above, below, left, and right. Fill in the rest.

Q. Sketch the ellipse $\frac{(x+3)^2}{49} + \frac{(y-5)^2}{625} = 1$, indicating the center, endpoints of the minor axis, and vertices.

A. This ellipse is the same as the one in the example in the earlier section "Determining Foci and Vertices of Ellipses." The center is at $(-3, 5)$, and the vertices are at $(-3, 30)$ and $(-3, -20)$. The minor axis is 14 units long; it goes 7 units on either side of the center, so the endpoints are at $(4, 5)$ and $(-10, 5)$. Mark the center and endpoints of the axes. Then fill in the rest (see Figure 10-6).

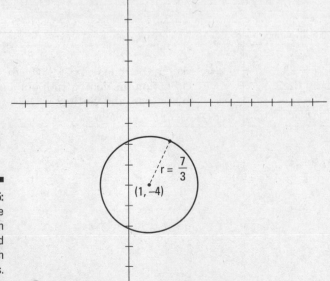

Figure 10-5: The circle stays in the third and fourth quadrants.

$r = \frac{7}{3}$

$(1, -4)$

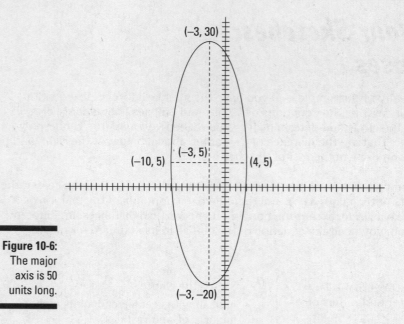

(−3, 30)

(−3, 5)

(−10, 5) - - - - - - - - (4, 5)

(−3, −20)

Figure 10-6:
The major
axis is 50
units long.

---

23. Sketch the circle $4x^2 + 4y^2 = 1$, indicating the center and radius on the graph.

24. Sketch the ellipse $x^2 + 9y^2 = 81$, indicating the center, endpoints of the minor axis, and vertices.

---

25. Sketch the circle $x^2 + y^2 - 10x + 2y = 23$, indicating the center and radius on the graph.

26. Sketch the ellipse $4x^2 + 25y^2 - 8x - 50y = 71$, indicating the center, endpoints of the minor axis, and vertices.

Hyperbola: Standard Equations and Foci

Of the four types of conic sections, the hyperbola is the only conic that seems a bit disconnected. The graph of a hyperbola is two separate curves seeming to face away from one another. The standard forms for the equation of hyperbolas are $\frac{(x-h)^2}{a^2} - \frac{(y-k)^2}{b^2} = 1$ or $\frac{(y-k)^2}{b^2} - \frac{(x-h)^2}{a^2} = 1$.

Notice that these formulas look just like the equation for the ellipse except for the minus sign between the two fractions.

Two forms of the standard equation exist; the form with the *x*-term in front is for hyperbolas that open to the left and right, and the form with the *y*-term in front is for hyperbolas that open upward and downward. The center of the hyperbola is the same old (h, k), as in the circles and ellipses.

You measure distances from the *foci* of a hyperbola to a point on the hyperbola. The *difference* between the distances (in the ellipse it's the *sum*) is always the same for any point on the hyperbola. Solve for the foci with $c^2 = a^2 + b^2$, and let $\pm c$ be the distance from the center to the foci, either vertically or horizontally (depending on the equation, which tells you whether the hyperbola opens up and down or left and right).

Q. Find the standard form of the hyperbola $16x^2 - 9y^2 = 144$. Then give the coordinates of the center and the coordinates of the foci.

A. $\frac{x^2}{9} - \frac{y^2}{16} = 1$; **center: (0, 0); foci: (–5, 0), (5, 0).** Divide each side of the equation by 144, and you get the standard form. The hyperbola opens left and right, because the *x* term appears first in the standard form. The center of the hyperbola is $(0, 0)$, the origin. To find the foci, solve for *c* with $c^2 = a^2 + b^2 = 9 + 16 = 25$. The value of *c* is ± 5. Counting 5 units to the left and right of the center, the coordinates of the foci are $(–5, 0)$ and $(5, 0)$.

Q. Find the standard form of the hyperbola $576(y - 5)^2 - 49(x - 3)^2 = 28{,}224$. Then give the coordinates of the center and the coordinates of the foci.

A. $\frac{(y-5)^2}{49} - \frac{(x-3)^2}{576} = 1$; **center: (3, 5); foci: (3, 30), (3, –20).** Divide each side of the equation by 28,224 (yes, the number is huge, but the fractions reduce very nicely) to get the standard form. The hyperbola opens upward and downward, because the *y* term appears first in the standard form. The center of the hyperbola is $(3, 5)$. To find the foci, solve for *c* with $c^2 = a^2 + b^2 = 49 + 576 = 625$. The value of *c* is ± 25. Counting 25 units upward and downward from the center, the coordinates of the foci are $(3, 30)$ and $(3, –20)$.

27. Find the standard form of the hyperbola $3x^2 - 18y^2 = 18$. Then give the coordinates of the center and the coordinates of the foci.

28. Find the standard form of the hyperbola $25y^2 - 144x^2 = 3{,}600$. Then give the coordinates of the center and the coordinates of the foci.

29. Find the standard form of the hyperbola $9x^2 - 4y^2 - 54x + 16y + 29 = 0$. Then give the coordinates of the center and the coordinates of the foci.

30. Find the standard form of the hyperbola $4y^2 - x^2 + 56y - 4x + 188 = 0$. Then give the coordinates of the center and the coordinates of the foci.

Determining the Asymptotes and Intercepts of Hyperbolas

Hyperbolas have two diagonal, intersecting asymptotes (literal guidelines) that help you with the general shape of the two curves. As with asymptotes of rational functions, these lines aren't really a part of the curve (see Chapter 8 for info on asymptotes). The asymptotes are helpful when graphing a hyperbola because they provide lines that the curves approach but don't cross.

Solve the equation $\dfrac{(x-h)^2}{a^2} = \dfrac{(y-k)^2}{b^2}$ for y to find the equations of the two asymptotes. You get this equation by setting the standard form equal to 0 instead of 1 and then finding the square root of each side, cross-multiplying, and finally writing the two linear equations of the asymptotes.

Hyperbolas have intercepts where they cross the x-axis or y-axis. If the center of the hyperbola is the origin, then the graph has x-intercepts or y-intercepts but not both. To solve for x-intercepts, set y equal to 0 and solve for x. To solve for y-intercepts, let x equal 0 and solve for y.

EXAMPLE

Q. Find the asymptotes and intercepts of the hyperbola $\dfrac{(x-6)^2}{4} - \dfrac{(y+3)^2}{9} = 1$.

A. **Asymptotes:** $y = \dfrac{3}{2}x - 12$, $y = -\dfrac{3}{2}x + 6$; **y-intercepts:**

$\left(0, -3 + 6\sqrt{2}\right), \left(0, -3 - 6\sqrt{2}\right)$;

x-intercepts: $\left(6 + 2\sqrt{2}, 0\right), \left(6 - 2\sqrt{2}, 0\right)$.

To find the asymptotes, solve the equation $\dfrac{(x-6)^2}{4} = \dfrac{(y+3)^2}{9}$ by taking the

square root of each side, cross-multiplying, and simplifying:

$$\pm\sqrt{\dfrac{(x-6)^2}{4}} = \sqrt{\dfrac{(y+3)^2}{9}}$$

$$\pm\dfrac{x-6}{2} = \dfrac{y+3}{3}$$

$$\pm\dfrac{3(x-6)}{2} = y+3$$

$$-3 \pm \left(\dfrac{3}{2}x - 9\right) = y$$

$$y = \dfrac{3}{2}x - 12 \text{ or } y = -\dfrac{3}{2}x + 6$$

You have to put the $\pm$ sign on only one side of the equation. (Putting it on both sides just creates four equations, two of which are repeats.) To find the y-intercepts, let x be 0 and solve for y:

$$\frac{(x-6)^2}{4} - \frac{(0+3)^2}{9} = 1$$

$$\frac{(x-6)^2}{4} - 1 = 1$$

$$\frac{(x-6)^2}{4} = 2$$

$$(x-6)^2 = 8$$

$$\sqrt{(x-6)^2} = \pm\sqrt{8}$$

$$x - 6 = \pm 2\sqrt{2}$$

$$x = 6 \pm 2\sqrt{2}$$

The two x-intercepts are approximately at $(8.828, 0)$ and $(3.172, 0)$. To solve for the x-intercepts, let the y in the equation be 0 and solve for x:

$$\frac{(0-6)^2}{4} - \frac{(y+3)^2}{9} = 1$$

$$9 - \frac{(y+3)^2}{9} = 1$$

$$8 = \frac{(y+3)^2}{9}$$

$$72 = (y+3)^2$$

$$\pm\sqrt{72} = \sqrt{(y+3)^2}$$

$$\pm 6\sqrt{2} = y + 3$$

$$y = -3 \pm 6\sqrt{2}$$

The y-intercepts are approximately at $(0, 5.485)$ and $(0, -11.485)$.

---

31. Find the asymptotes and intercepts of the hyperbola $\dfrac{x^2}{100} - \dfrac{y^2}{64} = 1$.

32. Find the asymptotes and intercepts of the hyperbola $\dfrac{y^2}{169} - \dfrac{x^2}{25} = 1$.

---

33. Find the asymptotes and intercepts of the hyperbola $\dfrac{(x-8)^2}{16} - \dfrac{(y+10)^2}{25} = 1$.

34. Find the asymptotes and intercepts of the hyperbola $\dfrac{y^2}{169} - \dfrac{(x+5)^2}{144} = 1$.

Sketching the Hyperbola

The intercepts, foci, asymptotes, and center of a hyperbola are all important characteristics of this conic, but you can make the actual sketch with just the center and a carefully placed rectangle:

- ✔ The center of the rectangle is the center of the hyperbola.

- ✔ The width of the rectangle is twice the square root of the value under the x's in the standard equation; the height of the rectangle is twice the square root of the number under the y's in the equation.

- ✔ The asymptotes are extended diagonals of the rectangle.

- ✔ The vertices of the hyperbola are at the midpoints of the rectangle's sides — either on the left and right sides or on the top and bottom, depending on which ways the hyperbola faces.

Q. Sketch the hyperbola $\dfrac{(x+4)^2}{4} - \dfrac{(y-3)^2}{64} = 1$.

A. The center of the hyperbola is (–4, 3). Insert a rectangle around that center by counting two units to the left and two units to the right of the center (these are the midpoints of the sides of the rectangle). Then count eight units up and eight units down from the center to find the other two midpoints. Lightly sketch in the rectangle. For the asymptotes, draw the diagonals of the rectangle and extend them. See Figure 10-7a for those first steps in the drawing. The hyperbola opens to the right and left, because the x-term comes first in the standard form. Starting with the midpoints on the sides of the rectangle, sketch in the hyperbola, having it get closer to the asymptote as it gets farther from the center. See Figure 10-7b for the entire sketch.

35. Sketch the hyperbola $\dfrac{(x-1)^2}{1} - \dfrac{(y+3)^2}{16} = 1$.

36. Sketch the hyperbola $\dfrac{(y-6)^2}{625} - \dfrac{(x-2)^2}{49} = 1$.

37. Sketch the hyperbola $x^2 - 9y^2 = 9$.

38. Sketch the hyperbola $9y^2 - 25x^2 + 54y + 150x = 369$.

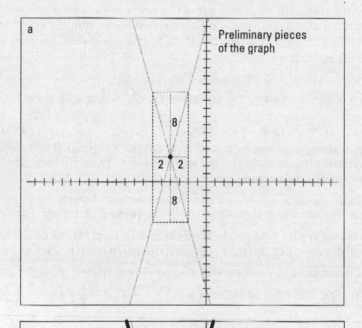

a   Preliminary pieces of the graph

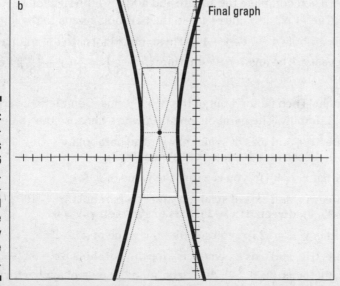

b   Final graph

Figure 10-7:
The rectangle is 4 units wide and 16 units high. The asymptotes and rectangle aren't really a part of the hyperbola.

Answers to Problems on Conic Sections

The following are the answers to the practice problems presented earlier in this chapter.

1 Write the equation of the parabola $x^2 - 6x - y + 4 = 0$ in standard form to determine its vertex and in which direction it opens. The answer is **equation: $(x - 3)^2 = y + 5$; vertex: (3, –5); opens upward.**

Complete the square on the left by moving the y and 4 to the right side and adding 9 to each side of the equation. Factor and simplify. The parabola opens upward, because the value of $4a$, the multiplier on the right, is +1.

2 Write the equation of the parabola $2x^2 + 8x + y + 3 = 0$ in standard form to determine its vertex and in which direction it opens. The answer is **equation: $(x + 2)^2 = -\frac{1}{2}(y - 5)$; vertex: (–2, 5); opens downward.**

Move the y and 3 to the right, and then factor 2 out of the two x terms to get $2(x^2 + 4x) = -y + 5$. Complete the square in the parentheses, and add 8 to the right side. Simplify and factor to get $2(x + 2)^2 = -1(y - 5)$. Divide each side by 2. The parabola opens downward, because the value of $4a$, the multiplier on the right, is $-\frac{1}{2}$.

3 Write the equation of the parabola $3y^2 - 12y - x + 9 = 0$ in standard form to determine its vertex and in which direction it opens. The answer is **equation: $(y - 2)^2 = \frac{1}{3}(x + 3)$; vertex: (–3, 2); opens to the right.**

Move the x and 9 to the right. Then factor 3 out of the two y terms to get $3(y^2 - 4y) = x - 9$. When you complete the square, you have to add 12 on the right. Simplifying and factoring, you get $3(y - 2)^2 = x + 3$. Divide each side by 3. The parabola opens to the right, because the value of $4a$, the multiplier on the right, is positive.

4 Write the equation of the parabola $y^2 - x + 4 = 0$ in standard form to determine its vertex and in which direction it opens. The answer is **equation: $y^2 = x - 4$; vertex: (4, 0); opens to the right.**

Add x and subtract 4 from each side, and you have the standard form. Think of the equation as $(y - 0)^2 = x - 4$. The parabola opens to the right, because the value of $4a$ is +1.

5 Write the equation of the parabola $y^2 + y + x = 0$ in standard form to determine its vertex and in which direction it opens. The answer is **equation: $\left(y + \frac{1}{2}\right)^2 = -1\left(x - \frac{1}{4}\right)$; vertex: $\left(\frac{1}{4}, -\frac{1}{2}\right)$; opens to the left.**

Subtract x from each side. Then complete the square and add $\frac{1}{4}$ to the right side. Factor –1 out of the terms on the right. The –1 is the value of $4a$, so the parabola opens to the left.

6 Write the equation of the parabola $4x^2 - x - y + 1 = 0$ in standard form to determine its vertex and in which direction it opens. The answer is **equation: $\left(x - \frac{1}{8}\right)^2 = \frac{1}{4}\left(y - \frac{15}{16}\right)$; vertex: $\left(\frac{1}{8}, \frac{15}{16}\right)$; opens upward.**

Move the y and 1 to the right. Then factor 4 out of the two x terms. Complete the square, adding $\frac{1}{16}$ to the right. Factor and simplify. The parabola opens upward, because the value of $4a$ is $\frac{1}{4}$.

7 Find the vertex, focus, directrix, and axis of symmetry for the parabola $x^2 = 16y$. The answer is **vertex: (0, 0); focus: (0, 4); directrix: $y = -4$; axis of symmetry: $x = 0$.**

When the equation is written $x^2 = 4(4)y$, you see that the value of a is 4.

8 Find the vertex, focus, directrix, and axis of symmetry for the parabola $y^2 = -100x$. The answer is **vertex: (0, 0); focus: (–25, 0); directrix: $x = 25$; axis of symmetry: $y = 0$.**

When the equation is written $y^2 = 4(-25)x$, you see that the value of a is –25.

9 Find the vertex, focus, directrix, and axis of symmetry for the parabola $8x^2 - 32x + 38 - y = 0$. The answer is **vertex: (2, 6); focus: $\left(2, 6\frac{1}{32}\right)$; directrix: $y = 5\frac{31}{32}$; axis of symmetry: $x = 2$.**

The standard form for this equation is

$$(x-2)^2 = \frac{1}{8}(y-6)$$

$$= 4\left(\frac{1}{32}\right)(y-6)$$

when a is equal to $\frac{1}{32}$. You get the standard form by using completing the square. Add y and subtract 38 from each side to get the equation $8x^2 - 32x = y - 38$. Factor 8 from the terms on the left, giving you $8(x^2 - 4x) = y - 38$. Complete the square, adding 32 to the right. $8(x^2 - 4x + 4) = y - 38 + 32$ factors into $8(x-2)^2 = y - 6$. Finally, divide each side by 8 to get the standard form.

10 Find the vertex, focus, directrix, and axis of symmetry for the parabola $y^2 - 2y + 4x + 13 = 0$. The answer is **vertex: (–3, 1); focus: (–4, 1); directrix: $x = -2$; axis of symmetry: $y = 1$.**

The standard form for this equation is $(y-1)^2 = -4(x+3)$. The value of a is –1. Use completing the square to solve for the standard form. First, subtract $4x$ and 13 from each side to get $y^2 - 2y = -4x - 13$. Now complete the square on the left, adding 1 to each side: $y^2 - 2y + 1 = -4x - 12$. Factor both sides, and you get $(y-1)^2 = -4(x+3)$. Divide each side by –4 to finalize the form.

11 Sketch the graph of $(x-4)^2 = 12(y+1)$. Label the vertex, focus, directrix, and axis of symmetry.

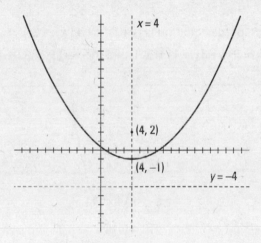

The value of a is 3, so the parabola is relatively wide.

12 Sketch the graph of $(y-1)^2 = 8(x-5)$. Label the vertex, focus, directrix, and axis of symmetry.

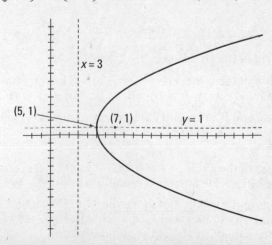

The value of a is 2, so the parabola widens out a bit.

13 Sketch the graph of $4y^2 + 24y + 16x + 33 = 0$. Label the vertex, focus, directrix, and axis of symmetry.

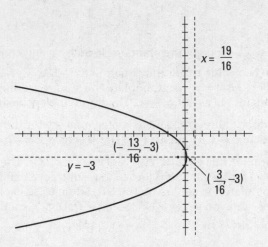

The standard form of the equation is $(y+3)^2 = -4\left(x - \dfrac{3}{16}\right)$.

14 Sketch the graph of $2x^2 + 8x + 11 = y$. Label the vertex, focus, directrix, and axis of symmetry.

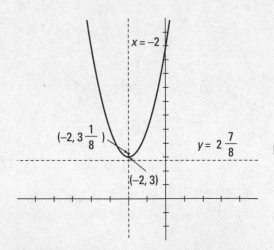

The standard form of the equation is $(x+2)^2 = \dfrac{1}{2}(y-3)$.

15 Write the equation of the circle $x^2 + y^2 - 6x + 8y = 11$ in standard form. Identify the center and radius of the circle. The answer is **equation: $(x - 3)^2 + (y + 4)^2 = 36$; center: (3, –4); radius: 6.**

Group the x terms and y terms and complete the square in each grouping: $(x^2 - 6x) + (y^2 + 8y) = 11$; $(x^2 - 6x + 9) + (y^2 + 8y + 16) = 11 + 9 + 16 = 36$. Factor the perfect square trinomials to get the standard form.

16 Write the equation of the circle $x^2 + y^2 + 2x - 10y + 17 = 0$ in standard form. Identify the center and radius of the circle. The answer is **equation: $(x + 1)^2 + (y - 5)^2 = 9$; center: (–1, 5); radius: 3.**

Group the x terms and y terms, and move the 17 to the right. Complete the square in each grouping: $(x^2 + 2x) + (y^2 - 10y) = -17$; $(x^2 + 2x + 1) + (y^2 - 10y + 25) = -17 + 1 + 25 = 9$. Factor the perfect square trinomials to get the standard form.

17 Write the equation of the ellipse $9x^2 + 4y^2 - 90x - 16y + 205 = 0$ in standard form. Identify the center and the lengths of the two axes. The answer is **equation:** $\dfrac{(x-5)^2}{4} + \dfrac{(y-2)^2}{9} = 1$; **center: (5, 2); major axis: 6 units; minor axis: 4 units.**

Group the x's together and the y's together; subtract 205 from each side. Factor 9 out of the x terms and 4 out of the y terms. You have $9(x^2 - 10x) + 4(y^2 - 4y) = -205$. Complete the square in each set of parentheses. Be sure to multiply what you insert into the parentheses by the factor outside. Then add that product to the other side: $9(x^2 - 10x + 25) + 4(y^2 - 4y + 4) = -205 + 225 + 16 = 36$. Factor the perfect square trinomials and divide each term by 36. The length of the major axis is twice the square root of 9: $2(3) = 6$. The length of the minor axis is twice the square root of 4: $2(2) = 4$.

18 Write the equation of the ellipse $x^2 + 16y^2 + 4x - 32y + 19 = 0$ in standard form. Identify the center and the lengths of the two axes. The answer is **equation:** $\dfrac{(x+2)^2}{1} + \dfrac{(y-1)^2}{1/16} = 1$; **center: (–2, 1); major axis: 2 units; minor axis: $\dfrac{1}{2}$ unit.**

Group the x's together and the y's together; subtract 19 from each side. Factor 16 out of the y terms. You have $(x^2 + 4x) + 16(y^2 - 2y) = -19$. Complete the square in each set of parentheses. Be sure to multiply what you insert into the second set of parentheses by the factor outside. Then add that product to the other side: $(x^2 + 4x + 4) + 16(y^2 - 2y + 1) = -19 + 4 + 16 = 1$. Factor the perfect square trinomials. Set the first term over 1. Multiply the numerator and denominator of the second term by $\dfrac{1}{16}$. The length of the major axis is twice the square root of 1: $2(1) = 2$. The length of the minor axis is twice the square root of $\dfrac{1}{16}$: $2\left(\dfrac{1}{4}\right) = \dfrac{1}{2}$.

19 Find the foci and vertices of the ellipse $\dfrac{(x+6)^2}{16} + \dfrac{(y-2)^2}{25} = 1$. The answer is **foci: (–6, 5), (–6, –1); vertices: (–6, 7), (–6, –3).**

Solving for c, $c^2 = 25 - 16 = 9$; $c = \pm 3$. The major axis runs vertically, because the larger denominator is under the y terms. Add and subtract 3 from the y-coordinate of the center, (–6, 2), to get the coordinates of the foci. Add and subtract 5, the value of b, from the y-coordinate of the center to get the coordinates of the vertices.

20 Find the foci and vertices of the ellipse $\dfrac{(x-1)^2}{625} + \dfrac{(y-7)^2}{576} = 1$. The answer is **foci: (8, 7), (–6, 7); vertices: (26, 7), (–24, 7).**

Solving for c, $c^2 = 625 - 576 = 49$; $c = +7$. The major axis runs horizontally, because the larger denominator is under the x terms. Add and subtract 7 from the x-coordinate of the center, (1, 7), to get the coordinates of the foci. Add and subtract 25, the value of a, from the x-coordinate of the center to get the coordinates of the vertices.

21 Find the foci and vertices of the ellipse $\dfrac{x^2}{169} + \dfrac{y^2}{144} = 1$. The answer is **foci: (5, 0), (–5, 0); vertices: (13, 0), (–13, 0).**

Solving for c, $c^2 = 169 - 144 = 25$; $c = \pm 5$. The major axis runs horizontally, because the larger denominator is under the x terms. Add and subtract 5 from the x-coordinate of the center, (0, 0), to get the coordinates of the foci. Add and subtract 13, the value of a, from the x-coordinate of the center to get the coordinates of the vertices.

22 Find the foci and vertices of the ellipse $100x^2 + 36y^2 = 3{,}600$. The answer is **foci: (0, 8), (0, –8); vertices: (0, 10), (0, –10).**

The standard form of the equation is $\dfrac{x^2}{36} + \dfrac{y^2}{100} = 1$. Solving for c, $c^2 = 100 - 36 = 64$; $c = \pm 8$. The major axis runs vertically, because the larger denominator is under the y terms. Add and subtract 8 from the y-coordinate of the center, (0, 0), to get the coordinates of the foci. Add and subtract 10, the value of b, from the y-coordinate of the center to get the coordinates of the vertices.

23 Sketch the circle $4x^2 + 4y^2 = 1$, indicating the center and radius on the graph.

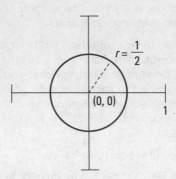

Divide each term in the equation by 4 to get the standard form, $x^2 + y^2 = \frac{1}{4}$. The radius is $\frac{1}{2}$.

24 Sketch the ellipse $x^2 + 9y^2 = 81$, indicating the center, endpoints of the minor axis, and vertices.

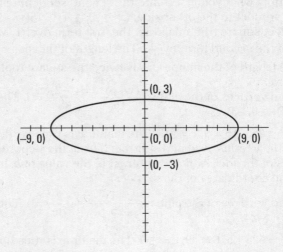

Divide each term in the equation by 81 to get the standard form, $\frac{x^2}{81} + \frac{y^2}{9} = 1$.

25 Sketch the circle $x^2 + y^2 - 10x + 2y = 23$, indicating the center and radius on the graph.

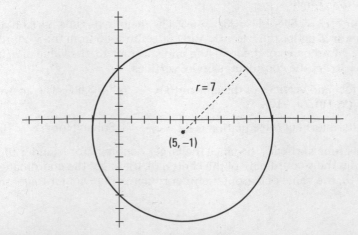

The standard form of the equation is $(x-5)^2 + (y+1)^2 = 49$.

26 Sketch the ellipse $4x^2 + 25y^2 - 8x - 50y = 71$, indicating the center, endpoints of the minor axis, and vertices.

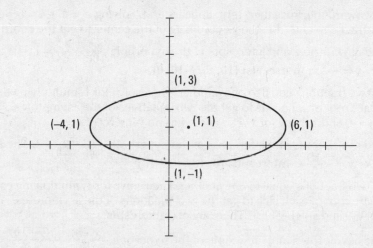

The standard form of the equation is $\dfrac{(x-1)^2}{25} + \dfrac{(y-1)^2}{4} = 1$.

27 Find the standard form of the hyperbola $3x^2 - 18y^2 = 18$. Then give the coordinates of the center and the coordinates of the foci. The answer is **equation:** $\dfrac{x^2}{6} - \dfrac{y^2}{1} = 1$; **center: (0, 0); foci:** $\left(\sqrt{7}, 0\right), \left(-\sqrt{7}, 0\right)$

Divide each term by 18 to get the standard form. The hyperbola opens left and right, because the x term appears first in the standard form. Solving $c^2 = 6 + 1 = 7$, you find that $c = \pm\sqrt{7}$. Add and subtract c to and from the x-coordinate of the center to get the coordinates of the foci.

28 Find the standard form of the hyperbola $25y^2 - 144x^2 = 3{,}600$. Then give the coordinates of the center and the coordinates of the foci. The answer is **equation:** $\dfrac{y^2}{144} - \dfrac{x^2}{25} = 1$; **center: (0, 0); foci: (0, 13), (0, -13).**

Divide each term by 3,600 to get the standard form. The hyperbola opens upward and downward, because the y term appears first in the standard form. Solving $c^2 = 144 + 25 = 169$, you find that $c = \pm 13$. Add and subtract c to and from the y-coordinate of the center to get the coordinates of the foci.

29 Find the standard form of the hyperbola $9x^2 - 4y^2 - 54x + 16y + 29 = 0$. Then give the coordinates of the center and the coordinates of the foci. The answer is **equation:** $\dfrac{(x-3)^2}{4} - \dfrac{(y-2)^2}{9} = 1$; **center: (3, 2); foci:** $\left(3 + \sqrt{13}, 2\right), \left(3 - \sqrt{13}, 2\right)$.

Group the x terms and y terms together, and subtract 29 from each side. Then factor 9 out of the x terms and -4 out of the y terms to get $9(x^2 - 6x) - 4(y^2 - 4y) = -29$. Completing the square in both sets of parentheses, you get $9(x^2 - 6x + 9) - 4(y^2 - 4y + 4) = -29 + 81 - 16 = 36$. Factor the trinomials, and then divide each term by 36 to get the standard form. The hyperbola opens left and right, because the x term appears first. Solving $c^2 = 4 + 9 = 13$, you get $c = \pm\sqrt{13}$. Add and subtract c to and from the x-coordinate of the center to get the coordinates of the foci.

30 Find the standard form of the hyperbola $4y^2 - x^2 + 56y - 4x + 188 = 0$. Then give the coordinates of the center and the coordinates of the foci. The answer is **equation:** $\dfrac{(y+7)^2}{1} - \dfrac{(x+2)^2}{4} = 1$; **center: (-2, -7); foci:** $\left(-2, -7 + \sqrt{5}\right), \left(-2, -7 - \sqrt{5}\right)$.

Group the y terms and x terms together, and subtract 188 from each side. Then factor 4 out of the y terms and –1 out of the x terms to get $4(y^2 + 14y) - (x^2 + 4x) = -188$. Completing the square in both sets of parentheses, you get $4(y^2 + 14y + 49) - (x^2 + 4x + 4) = -188 + 196 - 4 = 4$. Factor the trinomials, and then divide each term by 4 to get the standard form. The hyperbola opens upward and downward, because the y term appears first. Solving $c^2 = 1 + 4 = 5$, you find that $c = \pm\sqrt{5}$. Add and subtract c to and from the y-coordinate of the center to get the coordinates of the foci.

31 Find the asymptotes and intercepts of the hyperbola $\dfrac{x^2}{100} - \dfrac{y^2}{64} = 1$. The answer is **asymptotes:** $y = \dfrac{4}{5}x,\ y = -\dfrac{4}{5}x$; **intercepts: (10, 0), (–10, 0).**

Set the two fractions equal to one another. Then solve for y, multiplying each side by 64 and taking the square root of each side to get the two equations of the asymptotes. Find the x-intercepts by letting $y = 0$ and solving for x. The curve has no y-intercepts.

32 Find the asymptotes and intercepts of the hyperbola $\dfrac{y^2}{169} - \dfrac{x^2}{25} = 1$. The answer is **asymptotes:** $y = \dfrac{13}{5}x,\ y = -\dfrac{13}{5}x$; **intercepts: (0, 13), (0, –13).**

Set the two fractions equal to one another. Then solve for y, multiplying each side by 169 and taking the square root of each side to get the two equations of the asymptotes. Find the y-intercepts by letting $x = 0$ and solving for y. There are no x-intercepts.

33 Find the asymptotes and intercepts of the hyperbola $\dfrac{(x-8)^2}{16} - \dfrac{(y+10)^2}{25} = 1$. The answer is **asymptotes:** $y = \dfrac{5}{4}x - 20,\ y = -\dfrac{5}{4}x$; **intercepts:**
$$\left(0, -10+5\sqrt{3}\right), \left(0, -10-5\sqrt{3}\right), \left(8+4\sqrt{5}, 0\right), \left(8-4\sqrt{5}, 0\right).$$

Set the two fractions equal to one another. Then solve for y, multiplying each side by 25 and taking the square root of each side. Finally, subtract 10 from each side and simplify to get the two equations of the asymptotes. Determine the y-intercepts by letting $x = 0$ and solving for y. Find the x-intercepts by letting $y = 0$ and solving for x.

34 Find the asymptotes and intercepts of the hyperbola $\dfrac{y^2}{169} - \dfrac{(x+5)^2}{144} = 1$. The answer is **asymptotes:** $y = \dfrac{13}{12}x + \dfrac{65}{12},\ y = -\dfrac{13}{12}x - \dfrac{65}{12}$; **intercepts:** $\left(0, \dfrac{169}{12}\right), \left(0, -\dfrac{169}{12}\right)$.

Set the two fractions equal to one another. Then solve for y, multiplying each side by 169 and taking the square root of each side to get the two equations of the asymptotes. Find the y-intercepts by letting $x = 0$ and solving for y. There are no x-intercepts.

35 Sketch the hyperbola $\dfrac{(x-1)^2}{1} - \dfrac{(y+3)^2}{16} = 1$.

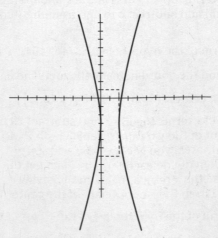

The center is at (1, –3), and the hyperbola opens left and right.

36 Sketch the hyperbola $\dfrac{(y-6)^2}{625} - \dfrac{(x-2)^2}{49} = 1$.

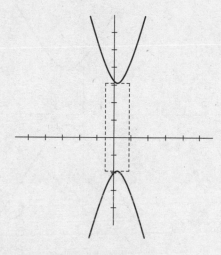

The center is at (2, 6), and the hyperbola opens upward and downward.

37 Sketch the hyperbola $x^2 - 9y^2 = 9$.

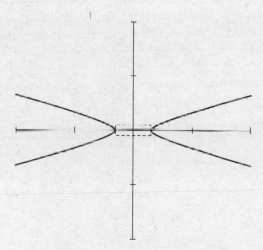

The standard form of the equation is $\dfrac{x^2}{9} - \dfrac{y^2}{1} = 1$. The center is at (0, 0), and the hyperbola opens left and right.

38 Sketch the hyperbola $9y^2 - 25x^2 - 54y + 150x = 369$.

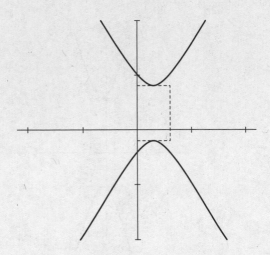

The standard form of the equation is $\dfrac{(y-3)^2}{25} - \dfrac{(x-3)^2}{9} = 1$. The center is at $(3, 3)$, and the hyperbola opens upward and downward.

Chapter 11

Solving Systems of Linear Equations

● ●

In This Chapter

▶ Solving systems of two and three linear equations using algebraic methods

▶ Using Cramer's rule to solve systems of two equations

▶ Applying systems of equations to breaking fractions apart

● ●

Linear equations have variables that don't exceed the power of one. For example, the equation $2x + 3y - z = 0$ is linear. A system of linear equations can contain any number of equations and any number of variables. But the only systems that have the possibility of having a unique solution are those where you have at least as many equations as variables. When looking for solutions of systems of equations, you try to get one numerical value for x, one for y, one for z, and so on. Another type of solution is a rule or generalization relating the values of the variables to one another.

In this chapter, you use substitution and elimination to solve the linear systems. You also use Cramer's rule as an alternative when fractional answers get nasty. Finally, you decompose fractions, which means breaking fractions into simpler fractions. Linear equations come in very handy when decomposition is desired (bet you never expected to read that decomposition is desirable!).

Solving Two Linear Equations Algebraically

A solution of a system of two linear equations consists of the values of x and y that make both of the equations true — at the same time. Graphically, the solution is the point where the two lines intersect. The two most frequently used methods for solving systems of linear equations are elimination and substitution:

✔ **Elimination (also called add-subtract):** This method involves adding the two equations together — or multiples of the two equations — so that in the sum, the coefficient on one of the variables becomes 0. That variable drops out (is eliminated), so you can solve for the other variable. Then you plug the solution back into one of the original equations and solve for the variable you eliminated.

✔ **Substitution:** This method has you set one of the equations equal to x or y. You can then substitute the equivalent of the variable from one equation for that variable into the other equation. You end up with a single-variable equation, which you can solve. Then plug that answer into one of the original equations and solve for the other variable.

You can use either method to solve linear systems, and you choose one over the other if a method seems to work better in a particular system (substitution works best if the coefficient on one of the variables is 1 or –1). The following examples show the same system of equations solved using both methods.

EXAMPLE

Q. Use elimination to solve for the common solution in the two equations: $x + 3y = 4$ and $2x + 5y = 5$.

A. $x = -5, y = 3$. Multiply each term in the first equation by –2 (you get $-2x - 6y = -8$) and then add the terms in the two equations together. You choose the number –2 as a multiplier because it makes the coefficient of the x term in the first equation equal to –2, while the coefficient on x in the second equation is 2. The numbers –2 and 2 are opposites, so adding the equations together eliminates the x term:

$$\begin{array}{r} -2x - 6y = -8 \\ 2x + 5y = 5 \\ \hline -y = -3 \end{array}$$

Now solve $-y = -3$ for y, and you get $y = 3$. Put 3 in for y in the first original equation, and you have $x + 3(3) = 4$; $x + 9 = 4$; $x = -5$. The solution is $x = -5, y = 3$, also written as the ordered pair (–5, 3). You can also solve for the x-value by putting the 3 into the second equation — you get the same result.

Q. Use substitution to solve for the common solution in the two equations: $x + 3y = 4$ and $2x + 5y = 5$.

A. $x = -5, y = 3$. To use substitution, select a variable in one of the equations with a coefficient of 1 or –1. The only variable that qualifies in this system is x in the first equation. Solve for x in terms of y in that equation. You get $x = 4 - 3y$. Substitute that equivalent of x into the second equation. The second equation becomes $2(4 - 3y) + 5y = 5$. Solve that equation for y: $8 - 6y + 5y = 5$; $8 - y = 5$; $-y = -3$; $y = 3$. That answer should look familiar. Substitute the 3 into $x + 3y = 4$ to get x: $x + 3(3) = 4$; $x + 9 = 4$; $x = -5$.

1. Solve for the common solution in the two equations: $5x - 3y = 7$ and $2x + 3y = 7$.

2. Solve for the common solution in the two equations: $8x - 3y = 41$ and $3x + 2y = 6$.

3. Solve for the common solution in the two equations: $4x + 5y = 11$ and $y = 2x + 5$.

4. Solve for the common solution in the two equations: $3x - 4y = 10$ and $x - 3y = 10$.

Using Cramer's Rule to Defeat Unruly Fractions

Cramer's rule — named for the mathematician Gabriel Cramer — is a way of solving a system of linear equations by focusing on just the coefficients and constants and ignoring (well, not really) the variables. You write the linear equations with the variables in exactly the same order each time, and then you can just leave the variables out of the computations. The technique is really rather slick, but you have to be organized and remember what you put where.

Cramer's rule works on systems of two linear equations — whether they're unruly or not. You can use elimination or substitution when the systems add together nicely or substitute one into the other (see the preceding section). This rule comes in handy when the coefficients of the variables are large or have no factors in common. Cramer is most appreciated when the answers are big, unwieldy fractions.

Essentially, Cramer's rule involves a bunch of cross-products. You create square arrangements of numbers, multiply across the diagonals, and find the difference to solve the systems of equations.

Writing a system of two linear equations as

$$a_1x + b_1y = c_1$$
$$a_2x + b_2y = c_2$$

let d represent the denominator, let n_x represent the x numerator, and let n_y represent the y numerator. The squares of numbers corresponding to these values are

$$d = \begin{vmatrix} a_1 & b_1 \\ a_2 & b_2 \end{vmatrix} \qquad n_x = \begin{vmatrix} c_1 & b_1 \\ c_2 & b_2 \end{vmatrix} \qquad n_y = \begin{vmatrix} a_1 & c_1 \\ a_2 & c_2 \end{vmatrix}$$

Use these squares to form the cross-products:

$$d = a_1b_2 - b_1a_2$$
$$n_x = c_1b_2 - b_1c_2$$
$$n_y = a_1c_2 - c_1a_2$$

The solution of the system of equations is then $x = \dfrac{n_x}{d} = \dfrac{c_1b_2 - b_1c_2}{a_1b_2 - b_1a_2}, y = \dfrac{n_y}{d} = \dfrac{a_1c_2 - c_1a_2}{a_1b_2 - b_1a_2}.$

The formulas for Cramer's rule aren't really as complicated as they may appear at first glance. When you're solving for x, you get the square of numbers for the numerator by replacing the x coefficients with the constants. And likewise, when you're solving for y, you get the square of numbers for the numerator by replacing the y coefficients with the constants.

Q. Solve the system of equations $12x + 13y = 10$ and $4x + 9y = 12$ using Cramer's rule.

A. $x = -\dfrac{33}{28}, y = \dfrac{13}{7}$. The three squares of numbers and their cross products are

$$d = \begin{vmatrix} 12 & 13 \\ 4 & 9 \end{vmatrix} = 12(9) - 13(4) = 108 - 52 = 56$$

$$n_x = \begin{vmatrix} 10 & 13 \\ 12 & 9 \end{vmatrix} = 10(9) - 13(12) = 90 - 156 = -66$$

$$n_y = \begin{vmatrix} 12 & 10 \\ 4 & 12 \end{vmatrix} = 12(12) - 10(4) = 144 - 40 = 104$$

To solve for x, divide n_x by d to get $\dfrac{-66}{56} = -\dfrac{33}{28}$. To solve for y, divide n_y by d to get $\dfrac{104}{56} = \dfrac{13}{7}$.

These fractions are perfectly good numbers. They just don't cooperate as well with the elimination or substitution methods.

---

5. Solve the system of equations $16x + 5y = 4$ and $13x + 3y = -1$ using Cramer's rule.

6. Solve the system of equations $5x + 7y = 11$ and $4x + 6y = 7$ using Cramer's rule.

7. Solve the system of equations $3x - 8y = 9$ and $2x - 5y = 4$ using Cramer's rule.

8. Solve the system of equations $6x - 9y = 8$ and $12x + 7y = 4$ using Cramer's rule.

A Third Variable: Upping the Systems to Three Linear Equations

When working with systems of equations, you can solve for one variable at a time. So, if a third linear equation comes along (bringing, of course, its variable z), well, three's a crowd. However, you can easily deal with all the variables as long as you address each in turn.

You solve systems of three (or more) linear equations using the elimination method ("Solving Two Linear Equations Algebraically," earlier in this chapter, explains how the basic method works):

1. **Starting with three equations, eliminate one variable to create two equations with the two remaining variables.**

   Pair the first equation with the second, the second with the third, or the first with the third to eliminate one of the variables. Then choose a different pairing and eliminate the same variable.

2. **From those two new equations, eliminate a second variable so you can solve for the one that remains.**

3. **Substitute back into the other equations to find the values of the other variables.**

   Plug the first variable you solved for into one of the two-variable equations you found in Step 1. Then solve for the third variable by plugging the known values into one of the original equations.

You may more easily understand this process with an example, so that's where I go next. (You can also solve systems of equations with matrices. See Chapter 14 for details.)

Q. Find the common solution of the system of equations $x + 5y - 2z = 2$, $4x + 3y + 2z = 2$, and $3x - 3y - 5z = 38$.

A. $x = 4$, $y = -2$, $z = -4$ — also written as the ordered triple $(4, -2, -4)$. You can choose to eliminate any of the three variables, but there's usually a good-better-best-worse-worst decision that can be made. In this problem, the best choice is to eliminate the x variable. The x variable has the only coefficient of 1 in all of the equations. You look for a 1 or –1 or for multiples of the same number in the coefficients of a single variable.

Do two pairings of elimination. Multiply the first equation by –4 and add it to the second equation:

$$\begin{aligned} -4x - 20y + 8z &= -8 \\ 4x + 3y + 2z &= 2 \\ \hline -17y + 10z &= -6 \end{aligned}$$

For the second pairing, multiply the first equation by –3 and add it to the third equation:

$$\begin{aligned} -3x - 15y + 6z &= -6 \\ 3x - 3y - 5z &= 38 \\ \hline -18y + z &= 32 \end{aligned}$$

Then add the two equations that result (after multiplying the second equation by –10 so you can eliminate the z's):

$$\begin{aligned} -17y + 10z &= -6 \\ 180y - 10z &= -320 \\ \hline 163y &= -326 \end{aligned}$$

Divide each side of the equation by 163 to get $y = -2$. Replace the y in $-18y + z = 32$ with the –2, and you get $-18(-2) + z = 32$; $36 + z = 32$; $z = -4$. Now take the values for y and z and put them into any of the original equations (I choose the first) to solve for x. You get $x + 5(-2) - 2(-4) = 2$; $x - 10 + 8 = 2$; $x - 2 = 2$; $x = 4$.

9. Find the common solution of the system of equations $3x + 4y - z = 7$, $2x - 3y + 3z = 5$, and $x + 5y - 2z = 0$.

10. Find the common solution of the system of equations $8x + 3y - 2z = -2$, $x - 3y + 4z = -13$, and $6x + 4y - z = -3$.

11. Find the common solution of the system of equations $x + y - 4z = 6$, $4x - 5y + 2z = 24$, and $3x - 8y - 3z = 5$.

12. Find the common solution of the system of equations $6x + y - 4z = 11$, $3x - y - 3z = 6$, and $9x + 4y + z = 5$.

A Line by Any Other Name: Writing Generalized Solution Rules

A system of linear equations has a single numerical solution (if one exists) or an infinite number of solutions (if one equation is a multiple or combination of others). Parallel lines never cross, so if the equations represent parallel lines, there's no solution.

A system such as $x - 3y = 7$ and $2x = 6y + 14$ has an infinite number of common solutions. The points $(4, -1)$, $(7, 0)$, and $(-2, -3)$ all work — to name a few. This situation occurs when one equation is actually a multiple of the other; the two equations are just two different ways of saying the same thing. Graphically, these equations represent the same line. A generalized solution, or a rule that creates all the possible solutions, is $(3y + 7, y)$. This answer is in a point form or ordered pair, (x, y), and it shows that you pick a number for y and then find the x by multiplying y by 3 and adding 7.

The same type of situation can also occur with three or more linear equations — where there's an infinite number of solutions, all bound by a relationship tied to one of the variables. You find these types of solutions when elimination completely wipes out one of the equations. When that happens, you solve for the variables in terms of the others.

Q. Solve the system of equations $x - 2y = 5$ and $8x = 16y + 40$

A. **$(2y + 5, y)$.** Rewrite the system using the same format for each equation so you can apply elimination. You get $x - 2y = 5$ and $8x - 16y = 40$. Multiply the terms in the first equation by -8 and add them to the second equation. You get

$$
\begin{array}{r}
-8x + 16y = -40 \\
\underline{8x + 16y = 40} \\
0 + 0 = 0
\end{array}
$$

Zero always equals 0. This is your signal that this system has an infinite number of solutions. Solve the first equation for x to get $x = 2y + 5$. You write your solution as $(2y + 5, y)$. The answer means that if you substitute any number for y, you get the corresponding value of x. All these solutions work in both of the original equations.

Q. Solve the system of equations $4x - 3y + z = 3$, $x - 4y - z = 15$, and $2x + 5y + 3z = -27$.

A. $\left(x, \dfrac{5x - 18}{7}, \dfrac{-13x - 33}{7}\right)$. Eliminate the z's. Add the first and second equations together to get $5x - 7y = 18$. Next, multiply the second equation by 3 and add it to the third:

$$
\begin{array}{r}
3x - 12y - 3z = 45 \\
\underline{2x + 5y + 3z = -27} \\
5x - 7y = 18
\end{array}
$$

You get the same equation as you did when adding the first two equations together. Combining the two resulting equations just gets you $0 = 0$. So, solve for either x or y in terms of the other, and then solve for z in terms of that same variable. Solving for y in terms of x, you get $5x - 18 = 7y$; $y = \dfrac{5x - 18}{7}$.

Now, to solve for z in terms of x, substitute for y in the first equation:

$$4x - 3\left(\frac{5x - 18}{7}\right) + z = 3$$

$$z = 3 - 4x + \frac{15x - 54}{7}$$

$$= \frac{21 - 28x}{7} + \frac{15x - 54}{7}$$

$$z = \frac{-13x - 33}{7}$$

The ordered triple representing the solution is $\left(x, \dfrac{5x - 18}{7}, \dfrac{-13x - 33}{7}\right)$. For instance, if you let $x = -2$, you get $y = -4$ and $z = -1$. The possibilities are endless.

13. Solve the system of equations: $x - y = 3$ and $2x = 2y + 6$.

14. Solve the system of equations: $3y - x = 7$ and $2x + 14 = 6y$.

15. Solve the system of equations: $x - y = 11$, $5x + 3z = 60$, and $5y + 3z = 5$.

16. Solve the system of equations: $x + 2y + z = 8$, $3x - 2y + 4z = 1$, and $x - 6y + 2z = -15$.

Decomposing Fractions Using Systems

To add or subtract fractions, you first find a common denominator. Then, when you perform the operation, you combine the numerators and leave the denominators the same.

You use the same principle for the reverse process: taking a fraction apart to see what was added or subtracted to get that result. The process is called *decomposition of fractions* or *partial fractions,* and it uses systems of equations to determine the solution. (Really, decomposition is a pretty nifty trick. I could introduce it to you with lots of letters representing constants, but I'd prefer not to muddy the waters! Take a look at the example to see how it's done.)

Q. Find the two fractions whose sum or difference is equal to $\dfrac{12x - 3}{2x^2 - 7x - 4}$.

A. $\dfrac{2}{2x + 1} + \dfrac{5}{x - 4}$. Here is how to get the answer, step by step:

1. Factor the denominator.

   The denominator of the fraction factors into $(2x + 1)(x - 4)$, so there are two fractions — one with a denominator of $2x + 1$ and another with a denominator of $x - 4$ — whose sum or difference is this fraction.

2. Set the original fraction equal to A over the first binomial plus B over the second binomial.

   Write $\dfrac{12x - 3}{2x^2 - 7x - 4} = \dfrac{A}{2x + 1} + \dfrac{B}{x - 4}$. The goal is to find out what A and B are equal to.

3. On the right, add the two fractions together by writing each with the common denominator; set the numerator equal to the numerator of the original problem.

Your common denominator is $(2x + 1)$ $(x - 4)$:

$$\frac{A}{2x+1} + \frac{B}{x-4} = \frac{A(x-4)}{(2x+1)(x-4)} + \frac{B(2x+1)}{(x-4)(2x+1)}$$
$$= \frac{A(x-4)+B(2x+1)}{(2x+1)(x-4)}$$

The numerator of the original problem must be equal to the numerator of the sum, so write $12x - 3 = A(x - 4) + B(2x + 1)$.

 You can do Step 3 by simply multiplying both sides of the equation by the least common denominator (use the factored denominator from Step 1).

4. On the right, distribute the A and the B; factor out the x's and group the remaining terms.

$$12x - 3 = Ax - 4A + 2Bx + B$$
$$12x - 3 = (A + 2B)x + (-4A + B)$$

5. Use these groupings to set up a system of equations; solve for A and B.

- The coefficient of x is equal to 12 on the left and $(A + 2B)$ on the right, so $12 = A + 2B$.

- The constants are equivalent to -3 and $(-4A + B)$, so $-3 = -4A + B$.

Solve the system of equations $12 = A + 2B$ and $-3 = -4A + B$ using elimination. Multiply the first equation by 4 and add it to the second equation:

$$
\begin{array}{rl}
48 = & 4A + 8B \\
-3 = & -4A + B \\
\hline
45 = & 9B \\
5 = & B
\end{array}
$$

Substituting $B = 5$ into the first equation, $12 = A + 2(5)$; $12 = A + 10$; $A = 2$.

6. Plug the A and B into your equation from Step 2.

Replacing the A with 2 and the B with 5 in the fractions, you get

$\dfrac{12x-3}{2x^2-7x-4} = \dfrac{2}{2x+1} + \dfrac{5}{x-4}$. The fraction is decomposed, but it certainly doesn't stink!

---

17. Find the two fractions whose sum or difference is equal to $\dfrac{10x+5}{2x^2+x-3}$.

18. Find the two fractions whose sum or difference is equal to $\dfrac{2x-22}{x^2-x-12}$.

Answers to Problems on Systems of Equations

The following are the answers to the practice problems presented earlier in this chapter.

1 Solve for the common solution in the two equations: $5x - 3y = 7$ and $2x + 3y = 7$. The answer is **$x = 2, y = 1$.**

The coefficients of the y terms are opposites of one another, so when you add the two equations together, you get $7x = 14$; $x = 2$. Replace the x with 2 in the first equation: $5(2) - 3y = 7$; $10 - 3y = 7$, $-3y = -3$; $y = 1$.

2 Solve for the common solution in the two equations: $8x - 3y = 41$ and $3x + 2y = 6$. The answer is **$x = 4, y = -3$.**

Multiply the terms in the first equation by 2 and the terms in the second equation by 3. As a result, you end up adding $-6y$ and $6y$ together, which eliminates the y terms when you add the two equations. You get $25x = 100$; $x = 4$. Replace the x with 4 in the second equation: $3(4) + 2y = 6$; $12 + 2y = 6$; $2y = -6$; $y = -3$.

3 Solve for the common solution in the two equations: $4x + 5y = 11$ and $y = 2x + 5$. The answer is **$x = -1, y = 3$.**

The second equation is already solved for y. Substitute the equivalent of y from the second equation into the first equation to get $4x + 5(2x + 5) = 11$. Distribute and simplify: $4x + 10x + 25 = 11$; $14x + 25 = 11$; $14x = -14$; $x = -1$. Replace the x with -1 in the second equation: $y = 2(-1) + 5 = 3$.

4 Solve for the common solution in the two equations: $3x - 4y = 10$ and $x - 3y = 10$. The answer is **$x = -2, y = -4$.**

Solve for x in the second equation to get $x = 3y + 10$. Substitute the equivalent of x into the first equation, and you get $3(3y + 10) - 4y = 10$. Distribute and simplify: $9y + 30 - 4y = 10$; $5y = -20$; $y = -4$. Now replace the y in the second equation to get $x - 3(-4) = 10$; $x + 12 = 10$; $x = -2$.

5 Solve the system of equations $16x + 5y = 4$ and $13x + 3y = -1$ using Cramer's rule. The answer is **$x = -1, y = 4$.**

$$x = \frac{n_x}{d} = \frac{\begin{vmatrix} 4 & 5 \\ -1 & 3 \end{vmatrix}}{\begin{vmatrix} 16 & 5 \\ 13 & 3 \end{vmatrix}} = \frac{4(3) - 5(-1)}{16(3) - 5(13)} = \frac{12 + 5}{48 - 65} = \frac{17}{-17} = -1$$

$$y = \frac{n_y}{d} = \frac{\begin{vmatrix} 16 & 4 \\ 13 & -1 \end{vmatrix}}{\begin{vmatrix} 16 & 5 \\ 13 & 3 \end{vmatrix}} = \frac{16(-1) - 4(13)}{16(3) - 5(13)} = \frac{-16 - 52}{48 - 65} = \frac{-68}{-17} = 4$$

6 Solve the system of equations $5x + 7y = 11$ and $4x + 6y = 7$ using Cramer's rule. The answer is $x = \dfrac{17}{2}, y = -\dfrac{9}{2}$.

$$x = \frac{n_x}{d} = \frac{\begin{vmatrix} 11 & 7 \\ 7 & 6 \end{vmatrix}}{\begin{vmatrix} 5 & 7 \\ 4 & 6 \end{vmatrix}} = \frac{11(6) - 7(7)}{5(6) - 7(4)} = \frac{66 - 49}{30 - 28} = \frac{17}{2}$$

$$y = \frac{n_y}{d} = \frac{\begin{vmatrix} 5 & 11 \\ 4 & 7 \end{vmatrix}}{\begin{vmatrix} 5 & 7 \\ 4 & 6 \end{vmatrix}} = \frac{5(7) - 11(4)}{5(6) - 7(4)} = \frac{35 - 44}{30 - 28} = \frac{-9}{2}$$

7 Solve the system of equations $3x - 8y = 9$ and $2x - 5y = 4$ using Cramer's rule. The answer is $x = -13, y = -6$.

$$x = \frac{n_x}{d} = \frac{\begin{vmatrix} 9 & -8 \\ 4 & -5 \end{vmatrix}}{\begin{vmatrix} 3 & -8 \\ 2 & -5 \end{vmatrix}} = \frac{9(-5) - (-8)(4)}{3(-5) - (-8)(2)} = \frac{-45 + 32}{-15 + 16} = \frac{-13}{1} = -13$$

$$y = \frac{n_y}{d} = \frac{\begin{vmatrix} 3 & 9 \\ 2 & 4 \end{vmatrix}}{\begin{vmatrix} 3 & -8 \\ 2 & -5 \end{vmatrix}} = \frac{3(4) - 9(2)}{3(-5) - (-8)(2)} = \frac{12 - 18}{-15 + 16} = \frac{-6}{1} = -6$$

8 Solve the system of equations $6x - 9y = 8$ and $12x + 7y = 4$ using Cramer's rule. The answer is $x = \dfrac{46}{75}, y = -\dfrac{12}{25}$.

$$x = \frac{n_x}{d} = \frac{\begin{vmatrix} 8 & -9 \\ 4 & 7 \end{vmatrix}}{\begin{vmatrix} 6 & -9 \\ 12 & 7 \end{vmatrix}} = \frac{8(7) - (-9)(4)}{6(7) - (-9)(12)} = \frac{56 + 36}{42 + 108} = \frac{92}{150} = \frac{46}{75}$$

$$y = \frac{n_y}{d} = \frac{\begin{vmatrix} 6 & 8 \\ 12 & 4 \end{vmatrix}}{\begin{vmatrix} 6 & -9 \\ 12 & 7 \end{vmatrix}} = \frac{6(4) - (8)(12)}{6(7) - (-9)(12)} = \frac{24 - 96}{42 + 108} = \frac{-72}{150} = -\frac{12}{25}$$

9 Find the common solution of the system of equations $3x + 4y - z = 7$, $2x - 3y + 3z = 5$, and $x + 5y - 2z = 0$. The answer is $x = 4$, $y = -2$, $z = -3$.

Eliminate x's by multiplying the third equation by -3 and adding it to the first equation; you get $-11y + 5z = 7$. Then eliminate x's in another combination by multiplying the original third equation by -2 and adding it to the second equation; you get $-13y + 7z = 5$. Use Cramer's rule on these two resulting equations (see the section "Using Cramer's Rule to Defeat Unruly Fractions"):

$$y = \frac{n_y}{d} = \frac{\begin{vmatrix} 7 & 5 \\ 5 & 7 \end{vmatrix}}{\begin{vmatrix} -11 & 5 \\ -13 & 7 \end{vmatrix}} = \frac{49 - 25}{-77 + 65} = \frac{24}{-12} = -2$$

$$z = \frac{n_z}{d} = \frac{\begin{vmatrix} -11 & 7 \\ -13 & 5 \end{vmatrix}}{\begin{vmatrix} -11 & 5 \\ -13 & 7 \end{vmatrix}} = \frac{-55 + 91}{-77 + 65} = \frac{36}{-12} = -3$$

Now substitute -2 for y and -3 for z in the original third equation to solve for x. You get $x + 5(-2) - 2(-3) = 0$; $x - 10 + 6 = 0$; $x - 4 = 0$; $x = 4$.

10 Find the common solution of the system of equations $8x + 3y - 2z = -2$, $x - 3y + 4z = -13$, and $6x + 4y - z = -3$. The answer is $x = -1$, $y = 0$, $z = -3$.

Eliminate z's by multiplying the first equation by 2 and adding it to the second equation to get $17x + 3y = -17$. Then eliminate z's in another combination by multiplying the third equation by 4 and adding it to the second equation; you get $25x + 13y = -25$. Use Cramer's rule on these two resulting equations:

$$x = \frac{n_x}{d} = \frac{\begin{vmatrix} -17 & 3 \\ -25 & 13 \end{vmatrix}}{\begin{vmatrix} 17 & 3 \\ 25 & 13 \end{vmatrix}} = \frac{-221 + 75}{221 - 75} = \frac{-146}{146} = -1$$

$$y = \frac{n_y}{d} = \frac{\begin{vmatrix} 17 & -17 \\ 25 & -25 \end{vmatrix}}{\begin{vmatrix} 17 & 3 \\ 25 & 13 \end{vmatrix}} = \frac{-425 + 425}{221 - 75} = \frac{0}{146} = 0$$

Now substitute $x = -1$ and $y = 0$ into the original third equation to get $6(-1) + 4(0) - z = -3$; $-6 - z = -3$; $-z = 3$; $z = -3$.

11 Find the common solution of the system of equations $x + y - 4z = 6$, $4x - 5y + 2z = 24$, and $3x - 8y - 3z = 5$. The answer is $x = 8$, $y = 2$, $z = 1$.

Eliminate y's by multiplying the first equation by 5 and adding it to the second equation to get $9x - 18z = 54$. Divide the terms in this equation by 9 to get $x - 2z = 6$, and then solve for x so you can do a substitution: $x = 2z + 6$. Now multiply the original first equation by 8 and add it to the third equation to get $11x - 35z = 53$. Substitute $2z + 6$ in for the x to get $11(2z + 6) - 35z = 53$; $22z + 66 - 35z = 53$; $-13z + 66 = 53$; $-13z = -13$; $z = 1$. The value of x is then $x = 2(1) + 6 = 8$. Replace the x with 8 and the z with 1 in the first original equation to get $8 + y - 4(1) = 6$; $y + 4 = 6$; $y = 2$.

12 Find the common solution of the system of equations $6x + y - 4z = 11$, $3x - y - 3z = 6$, and $9x + 4y + z = 5$. The answer is $x = \frac{1}{3}$; $y = 1$; $z = -2$.

Eliminate y's by adding the first and second equations together to get $9x - 7z = 17$. Now multiply the second equation by 4 and add it to the third equation to get $21x - 11z = 29$. Use Cramer's rule on these two resulting equations:

$$x = \frac{n_x}{d} = \frac{\begin{vmatrix} 17 & -7 \\ 29 & -11 \end{vmatrix}}{\begin{vmatrix} 9 & -7 \\ 21 & -11 \end{vmatrix}} = \frac{-187 + 203}{-99 + 147} = \frac{16}{48} = \frac{1}{3}$$

$$z = \frac{n_z}{d} = \frac{\begin{vmatrix} 9 & 17 \\ 21 & 29 \end{vmatrix}}{\begin{vmatrix} 9 & -7 \\ 21 & -11 \end{vmatrix}} = \frac{261 - 357}{-99 + 147} = \frac{-96}{48} = -2$$

Substitute the values of x and z back into the first equation to get $6\left(\frac{1}{3}\right) + y - 4(-2) = 11$; $2 + y + 8 = 11$; $y = 1$.

13 Solve the system of equations: $x - y = 3$ and $2x = 2y + 6$. The answer is $(x, x - 3)$ or $(y + 3, y)$.

Rewrite the second equation as $2x - 2y = 6$. Multiplying the terms in the first equation by -2 and adding them to the second equation, you get $0 = 0$. This system has an infinite number of solutions, so write the general rule for the solutions by solving for either y or x. Solving the first equation for y, you get $y = x - 3$, so the general rule as an ordered pair is $(x, x - 3)$; after you choose a value for x, you can subtract 3 from that number to get y. For instance, $(5, 2)$, $(3, 0)$, and $(-4, -7)$ are all ordered pairs using this rule. If you, instead, chose to solve for x, you get $x = y + 3$, and the general rule as an ordered pair is $(y + 3, y)$. Letting $y = 2$, 0, or -7, you get the same ordered pairs listed before.

14 Solve the system of equations: $3y - x = 7$ and $2x + 14 = 6y$. The answer is $(3y - 7, y)$ or $\left(x, \frac{x + 7}{3}\right)$.

Rewrite the second equation as $-6y + 2x = -14$. Then multiply the terms in the first equation by 2 and add the equations together. You get $0 = 0$, an indication that multiple solutions exist. Solving the first equation for x, you get $x = 3y - 7$, so the ordered pair for the general solution is $(3y - 7, y)$. After you choose a value for y, you can solve for x using the rule in the first coordinate position. For instance, if $y = 6$, $x = 11$. If you choose to solve for y instead of x, you see that your general rule has a fraction. That's not bad — it's just not as nice to work with.

15 Solve the system of equations: $x - y = 11$, $5x + 3z = 60$, and $5y + 3z = 5$. The answer is $\left(x, x - 11, \frac{60 - 5x}{3}\right)$ or $\left(y + 11, y, \frac{5 - 5y}{3}\right)$ or $\left(\frac{60 - 3z}{5}, \frac{5 - 3z}{5}, z\right)$.

Multiply the terms in the third equation by -1 and add them to the second equation to get $5x - 5y = 55$. Divide each term by 5, and the result is the same as the first equation, so this system has an infinite number of solutions. Solving the first equation for y yields $y = x - 11$. Replace the y in the third equation with $x - 11$, and you get $5(x - 11) + 3z = 5$; $5x - 55 + 3z = 5$; $5x + 3z = 60$. Solve for z by subtracting $5x$ from each side and dividing each term by 3. After you choose a value for x, you can solve for y and z. For instance, if $x = 15$, then $y = 4$, and $z = -5$. You can see that the answers work in the other general solutions, too.

16 Solve the system of equations: $x + 2y + z = 8$, $3x - 2y + 4z = 1$, and $x - 6y + 2z = -15$. The answer is $\left(\dfrac{9-5z}{4}, \dfrac{23+z}{8}, z\right)$ or $\left(x, \dfrac{31-x}{10}, \dfrac{9-4x}{5}\right)$ or $(31 - 10y, y, 8y - 23)$.

Add the first two equations together to get $4x + 5z = 9$. Then multiply the terms in the first equation by 3 and add them to the third equation. Once again, you get $4x + 5z = 9$. This system has an infinite number of solutions. Solve for x by subtracting $5z$ from each side and dividing by 4. Then substitute this equivalent of x into the first equation and simplify (first multiply each term by 4 and then combine like terms): $\dfrac{9-5z}{4} + 2y + z = 8$, $9 - 5z + 8y + 4z = 32$, $-z + 8y = 23$.

Now add z to each side and divide by 8: $8y = 23 + z$, $y = \dfrac{23+z}{8}$. Letting $z = 5$, you get $x = -4$ and $y = \dfrac{7}{2}$. You can also write the general solution in terms of x or y, as shown.

17 Find the two fractions whose sum or difference is equal to $\dfrac{10x+5}{2x^2+x-3}$. The answer is $\dfrac{4}{2x+3} + \dfrac{3}{x-1}$.

The denominator of the fraction factors into $(2x + 3)(x - 1)$. Write the equation involving the original fraction and the sum of the two fractions with unknown numerators: $\dfrac{10x+5}{2x^2+x-3} = \dfrac{A}{2x+3} + \dfrac{B}{x-1}$. Multiply each side of the equation by the common denominator, and you're left with just the numerators: $\cancel{(2x+3)(x-1)}\dfrac{10x+5}{\cancel{(2x+3)(x-1)}} = \dfrac{A}{\cancel{(2x+3)}}\cancel{(2x+3)}(x-1) +$

$\dfrac{B}{\cancel{x-1}}(2x+3)\cancel{(x-1)}$ $10x+5 = A(x-1) + B(2x+3)$.

Simplify the equation on the right by distributing, rearranging, and factoring:

$$10x + 5 = Ax - A + 2Bx + 3B$$
$$= Ax + 2Bx - A + 3B$$
$$= (A + 2B)x + (-A + 3B)$$

Equating the coefficients of x, you get $10 = A + 2B$. Equating the constants, you get $5 = -A + 3B$. Add the first of these new equations to the second to get $15 = 5B$; $B = 3$. Replace the B in either of the two new equations and solve for A: $A = 4$.

18 Find the two fractions whose sum or difference is equal to $\dfrac{2x-22}{x^2-x-12}$. The answer is $\dfrac{-2}{x-4} + \dfrac{4}{x+3}$.

The denominator of the fraction factors into $(x - 4)(x + 3)$. Write the equation involving the original fraction and the sum of the two fractions with unknown numerators: $\dfrac{2x-22}{x^2-x-12} = \dfrac{A}{x-4} + \dfrac{B}{x+3}$. Multiply each side of the equation by the common denominator, and you're left with just the numerators: $\cancel{(x-4)(x+3)}\dfrac{2x-22}{\cancel{(x-4)(x+3)}} = \dfrac{A}{\cancel{(x-4)}}\cancel{(x-4)}(x+3)$

$+ \dfrac{B}{\cancel{(x+3)}}(x-4)\cancel{(x+3)}$.

Simplify the equation on the right by distributing, rearranging, and factoring:

$$2x - 22 = Ax + 3A + Bx - 4B$$
$$= Ax + Bx + 3A - 4B$$
$$= (A + B)x + (3A - 4B)$$

Equating the coefficients of x, you get $2 = A + B$. Equating the constants, you get $-22 = 3A - 4B$. Multiply the first of these new equations by 4 and add it to the second to get $-14 = 7A$; $A = -2$. Substitute -2 for A in the first equation, and you get that $B = 4$.

Chapter 12

Solving Systems of Nonlinear Equations and Inequalities

The graphs of a line and a parabola may intersect in two points or one point or not at all. So, the solution of a system of equations involving both linear and quadratic equations, whose graphs are a line and a parabola, may have two solutions, one solution, or none at all. (Don't you just love ambiguity?) The number of possible solutions increases with the powers on the variables of the equations.

In this chapter, you deal with the multiple solutions you find when working with systems that mix lines, parabolas, and circles. You also dabble with systems of inequalities to find a *feasible region* or area where all the points in the area are solutions of the system. Read on to find the methods to this madness.

Finding the Intersections of Lines and Parabolas

A line can cut through a parabola in two points, or it may just be tangent to the parabola and touch it at one point. And then, sadly, a line and a parabola may never meet. When solving systems of equations involving lines and parabolas, you usually use the substitution method — solving for x or y in the equation of the line and substituting into the equation of the parabola.

Sometimes the equations lend themselves to elimination — when adding the equations (or multiples of the equations) together eliminates one of the variables entirely because its coefficient becomes 0. Elimination works only occasionally, but substitution always works. (For more information on the basics of substitution and elimination, see Chapter 11.)

Q. Find the common solution(s) in the equations $y = -5x^2 + 12x + 3$ and $8x + y = 18$.

A. The points of intersection are **(1, 10), (3, –6)**. Here's another way to write this solution: When $x = 1$, $y = 10$, and when $x = 3$, $y = -6$. To find these solutions, rewrite the equation of the line as $y = 18 - 8x$. Replace the y in the equation of the parabola with its equivalent to get $18 - 8x = -5x^2 + 12x + 3$. Move all the terms to the left and combine like terms, giving you $5x^2 - 20x + 15 = 0$. Divide each term by 5 and then factor, which gives you the equation $5(x^2 - 4x + 3) = 5(x - 3)(x - 1) = 0$. Using the multiplication property of zero (in order for a product to equal 0, one of the factors must be 0), you know that $x = 3$ or $x = 1$. Substitute those values back into the equation of the line to get the corresponding y-values.

Always substitute back into the equation with the lower exponents. You can avoid creating extraneous solutions.

Q. Find the common solution(s) in the equations $y = x^2 - 4x$ and $2x + y + 1 = 0$

A. **(1, –3).** Solve for y in the equation of the line to get $y = -2x - 1$. Substitute this value into the equation of the parabola to get $-2x - 1 = x^2 - 4x$. Moving the terms to the right and simplifying, $0 = x^2 - 2x + 1 = (x - 1)^2$. The only solution is $x = 1$. Replacing x with 1 in the equation of the line, you find that $y = -3$. The line is tangent to the parabola at the point of intersection, which is why this problem has only one solution.

1. Find the common solution(s) in the equations $y = x^2 + 4x + 7$ and $3x - y + 9 = 0$.

2. Find the common solution(s) in the equations $y = 4x^2 - 8x - 3$ and $4x + y = 5$.

3. Find the common solution(s) in the equations $x = 2y^2 - y$ and $x + 3y = 12$.

4. Find the common solution(s) in the equations $x = y^2 + 5y + 6$ and $x - 3y = 6$.

Crossing Curves: Finding the Intersections of Parabolas and Circles

When a parabola and circle intersect, the possibilities for their meeting are many and varied. The two curves can intersect in as many as four different points, or maybe three, or just two or even just one point. Keep your options open and be alert for as many common solutions as possible (right — up to four). And, yes, the system may have no solution at all. The curves may miss each other completely. I guess some things just aren't meant to be.

For these problems, you usually turn to substitution. However, you don't have to set one of the equations equal to x or y by itself. You may solve an equation for $4x$ or $(y-3)^2$ or some other term that appears in the other equation. As long as the terms match, you can swap one value for the other.

Q. Find the common solutions of the circle $(x-2)^2 + (y-2)^2 = 4$ and the parabola $2y = x^2 - 4x + 4$.

A. **(0, 2), (2, 0), (4, 2).** Rewrite the equation of the parabola as $2y = (x-2)^2$. Next, replace the $(x-2)^2$ term in the first equation with $2y$ and simplify: $2y + (y-2)^2 = 4$; $2y + y^2 - 4y + 4 = 4$; $y^2 - 2y = 0$. Factor the terms on the left to get $y(y-2) = 0$. So, $y = 0$ or $y = 2$. Letting $y = 0$ in the equation of the parabola, you get $2(0) = x^2 - 4x + 4$, or $0 = (x-2)^2$. So, when $y = 0$, $x = 2$. Next, let $y = 2$ in the parabola equation. You get $2(2) = x^2 - 4x + 4$; $4 = x^2 - 4x + 4$; $0 = x^2 - 4x$. Factoring, $0 = x(x-4)$, so $x = 0$ or $x = 4$. When $y = 2$, $x = 0$ or 4.

Q. Find the common solutions of $x^2 + y^2 = 100$ and $y^2 + 6x = 100$.

A. **(0, 10), (0, –10), (6, 8), (6, –8).** Solving the second equation for y^2, you get $y^2 = 100 - 6x$. Replace the y^2 in the first equation with its equivalent to get $x^2 + 100 - 6x = 100$. Simplifying and factoring, the equation becomes $x^2 - 6x = x(x-6) = 0$. So $x = 0$ or $x = 6$. Replacing x with 0 in the equation of the parabola, $y^2 = 100$; $y = \pm 10$. Replacing x with 6 in the equation of the parabola, $y^2 + 36 = 100$; $y^2 = 64$; $y = \pm 8$.

5. Find the common solutions of $x^2 + y^2 = 25$ and $x^2 + 4y = 25$.

6. Find the common solutions of $x^2 + y^2 = 9$ and $5x^2 - 6y = 18$.

7. Find the common solutions of $x^2 + y^2 = 5$ and $y^2 = -4x$.

8. Find the common solutions of $x^2 + y^2 = 9$ and $y = x^2 + 11$.

Appealing to a Higher Power: Dealing with Exponential Systems

You can solve systems of exponential equations algebraically when the bases of the exponential terms are the same number or when *obvious* (don't you hate that word in mathematics?) solutions pop out because of the simple natures of the equations involved. If the bases match, you can simply set the exponents equal to each other. When an algebraic solution isn't available, then a good graphing calculator or computer program can find the solution — which usually includes lots of decimal values and/or logarithmic functions.

This section deals with the types of problems that you can solve algebraically (or *simply*). Of course, the most simple is just plugging in a number that you're pretty sure works. But that method can be time consuming if you have to plug away a lot — reserve it for the sure thing. Refer to Chapter 9 if you need to know more about solving equations that involve exponential terms.

Q. Find the common solutions of $y = 2^{x^2+6x}$ and $y = 16^{x+2}$.

A. $(2, 65{,}536)$, $\left(-4, \dfrac{1}{256}\right)$. You want the bases to match, so first change the exponential term in the second equation to a power of 2. It becomes $y = (2^4)^{x+2} = 2^{4x+8}$. Setting the two y-values of the two different equations equal to one another, you get $2^{x^2+6x} = 2^{4x+8}$. Now set the two exponents equal to each other: $x^2 + 6x = 4x + 8$. Moving all the terms to the left and factoring, $x^2 + 2x - 8 = (x + 4)(x - 2) = 0$. The solutions of this quadratic equation are $x = -4$ or $x = 2$. Replace the x with -4 in either of the original equations, and you get $y = \dfrac{1}{256}$. Replace x with 2 in either equation, and you get $y = 65{,}536$.

Q. Find the common solutions of $y = 3^{x+1}$ and $y = 2x + 3$

A. $(0, 3)$, $(-1, 1)$. A graphing calculator would show you an exponential curve rising from left to right and a line appearing to cut through the curve in two places near the y-axis. You'd have to zoom in closely to see the two points of intersection. These equations were chosen carefully so that the answers are integers. If you evaluate the two functions for a few values, you can determine the solutions with minimal computation. Let $x = 0$ in the first equation, and you get $y = 3^{0+1} = 3$. Let $x = 0$ in the second equation, and you get $y = 2(0) + 3 = 3$. A solution! Let $x = -1$ in the first equation, and you get $y = 3^{-1+1} = 3^0 = 1$. Let $x = -1$ in the second equation, and you get $y = 2(-1) + 3 = -2 + 3 = 1$. These are the only two solutions.

9. Find the common solution(s) of $y = 3^{x-1}$ and $y = 9^x$.

10. Find the common solution(s) of $y = 8^{2-x}$ and $y = 4^{x^2-x}$.

11. Find the common solution(s) of $y = 2^x$ and $y = 1 - x$.

12. Find the common solution(s) of $y = e^{x^2}$ and $y = e$.

Solving Systems of Inequalities

An inequality such as $x > 2$ has an infinite number of solutions. Any number bigger than 2 is a solution. The inequality $y > 4$ also has an infinite number of solutions. Put the two inequalities together and solve for all the ordered pairs (x, y) that make both inequalities true at the same time, and you have lots and lots of points, such as $(3, 5)$, $(4, 6)$, $(3, 8)$, and so on.

The best way to represent all the solutions of this system of inequalities is with a picture. You show a graph that has the upper-right corner all shaded in — everything to the right of $x = 2$ and everything above $y = 4$. This drawing method is essentially how you solve any system of inequalities: Shade in everything (all the points) shared by the inequalities in the system.

Here's how the process works:

1. **If the inequalities aren't already in their proper forms, rewrite them so you can more easily graph the curves (yes, straight lines are types of curves).**

   This step often means isolating the y (or x) on one side of the inequality so you can plot some points (solving for y can give you slope-intercept form of a line — see Chapter 4). If you're working with a circle or ellipse, graphing may be easier if you write the inequality in its standard form (see the equations in Chapter 10 for details). Remember to reverse the inequality sign whenever you divide by a negative number.

2. **Temporarily let the inequality signs be equal signs, and graph the curves you create.**

   Use dashed lines if the equation says $>$ or $<$; use solid lines if it includes $\geq$ or $\leq$.

3. **Try a test point to determine whether a region contains solutions to one of the inequalities.**

   Select a *test point*, a random point that's clearly on one side of a curve or the other (I suggest something simple, like the origin). Replace the x and y in the chosen inequality with the test coordinates and see whether the statement is true.

4. **Lightly shade in the region that makes the statement true; repeat steps 3 and 4 for the other curves.**

   Your solution for the system of equations is where the shading overlaps.

Q. Determine the solution of the system of inequalities $y \geq x + 4$ and $x + y + 2 > 0$.

A. Sketch in the graphs of the lines $y = x + 4$ and $x + y + 2 = 0$. Use a solid line for the first line, because the inequality has *greater than or equal to*, which means you have to include all the points on the line, too. The graph of the second line should be a dotted or dashed line to show that you don't include the points on the line in the solution. Figure 12-1 shows the graphs of the lines and their point of intersection $(-3, 1)$. You find that point of intersection using either elimination or substitution (see Chapter 11 if you need more info on finding solutions of systems of linear equations).

The two intersecting lines divide the graphing plane into four different regions. You need to determine which region

contains the solutions of the system. All the points that satisfy the inequality $y \geq x + 4$ are to the left of the line. You can tell that by just trying a test point, such as $(0, 0)$, which is clearly below the line. Replace the x and y in the inequality with 0s, and you get $0 \geq 0 + 4$. That's not true, so the point $(0, 0)$ isn't in the solution of the inequality. The answers must be on the other side of the line. Lightly shade in the area above that line. Trying that same test point in the second inequality, you get $0 + 0 + 2 > 0$, which is true. So the point $(0, 0)$ is a solution of the second inequality — as are all the points to the right of the line. Shade in the right side of that second line. The two shaded areas overlap in the upper region of the graph, and the shaded area is darker there. All the points in that region are solutions of the system of inequalities.

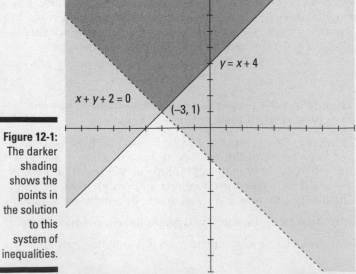

Figure 12-1:
The darker shading shows the points in the solution to this system of inequalities.

13. Determine the solution of the system of inequalities: $x + y \leq 10$ and $2x + y > 18$

14. Determine the solution of the system of inequalities: $3x + 5y \leq 15$, $x \geq 0$, and $y \geq 0$

15. Determine the solution of the system of inequalities: $2x - y \leq 22$, $x \leq 2$, and $y \leq 0$

16. Determine the solution of the system of inequalities: $y \geq 2x^2 - 3x + 4$ and $x \geq y - 4$

Answers to Problems on Solving Systems of Nonlinear Equations and Inequalities

The following are the answers to the practice problems presented earlier in this chapter.

1 Find the common solution(s) in the equations $y = x^2 + 4x + 7$ and $3x - y + 9 = 0$. The answer is **(–2, 3), (1, 12)**.

Solve for y in the second equation (you get $y = 3x + 9$), and substitute that into the equation of the parabola: $3x + 9 = x^2 + 4x + 7$. Move all the terms to the right and factor the equation: $0 = x^2 + x - 2 = (x + 2)(x - 1)$. So, $x = -2$ or 1. Letting $x = -2$ in the equation of the line, $3(-2) - y + 9 = 0$; $-6 - y + 9 = 0$; $-y = -3$; $y = 3$. And when $x = 1$ in the equation of the line, $3(1) - y + 9 = 0$; $3 - y + 9 = 0$; $-y = -12$; $y = 12$.

REMEMBER

When solving for the second coordinate in the solution of a system of equations, use the simpler equation — the one with the smaller exponents — to avoid introducing extraneous solutions.

2 Find the common solution(s) in the equations $y = 4x^2 - 8x - 3$ and $4x + y = 5$. The answer is **(–1, 9), (2, –3)**.

Solve for y in the second equation (you get $y = 5 - 4x$) and substitute the equivalent of y into the equation of the parabola: $5 - 4x = 4x^2 - 8x - 3$. Move all the terms to the right and factor the equation: $0 = 4x^2 - 4x - 8 = 4(x^2 - x - 2) = 4(x + 1)(x - 2)$. Using the multiplication property of zero, you find that $x = -1$ or $x = 2$. When $x = -1$ in the equation of the line, $4(-1) + y = 5$; $-4 + y = 5$; $y = 9$. And substituting $x = 2$ in the equation of the line, $4(2) + y = 5$; $8 + y = 5$; $y = -3$.

3 Find the common solution(s) in the equations $x = 2y^2 - y$ and $x + 3y = 12$. The answer is **(21, –3), (6, 2)**.

Solve for x in the second equation (you get $x = 12 - 3y$) and substitute that value into the equation of the parabola to get $12 - 3y = 2y^2 - y$. Move all the terms to the right and factor the equation: $0 = 2y^2 + 2y - 12 = 2(y^2 + y - 6) = 2(y + 3)(y - 2)$. The solutions are $y = -3$ or $y = 2$. Substitute –3 for y in the equation of the line, and you get $x + 3(-3) = 12$; $x - 9 = 12$; $x = 21$. Next, substituting 2 for y, $x + 3(2) = 12$; $x + 6 = 12$; $x = 6$.

4 Find the common solution(s) in the equations $x = y^2 + 5y + 6$ and $x - 3y = 6$. The answer is **(6, 0), (0, –2)**.

Solve for x in the second equation (you get $x = 3y + 6$), and substitute that value into the equation of the parabola to get $3y + 6 = y^2 + 5y + 6$. Move all the terms to the right and factor the equation: $0 = y^2 + 2y = y(y + 2)$. The solutions are $y = 0$ or $y = -2$. Substitute 0 for y in the equation of the line, and you determine that $x - 3(0) = 6$; $x - 0 = 6$; $x = 6$. Next, substituting –2 for y, $x - 3(-2) = 6$; $x + 6 = 6$; $x = 0$.

5 Find the common solutions of $x^2 + y^2 = 25$ and $x^2 + 4y = 25$. The answer is **(5, 0), (–5, 0), (3, 4), (–3, 4)**.

Solve the second equation for x^2 (you get $x^2 = 25 - 4y$) and replace the x^2 in the first equation with its equivalent. The new equation reads $25 - 4y + y^2 = 25$. Simplifying, you get $y^2 - 4y = 0$. This equation factors into $y(y - 4) = 0$. The two solutions are $y = 0$ and $y = 4$. Go back to the second equation, the equation of the parabola, because it has only one squared term (it has lower exponents, so choosing this equation lets you avoid extraneous solutions). Replace the y in that equation with 0 to get $x^2 + 4(0) = 25$; $x^2 = 25$. That equation has two solutions: $x = 5$ or

$x = -5$. Now, going back and replacing the y with 4 in the equation of the parabola, $x^2 + 4(4) = 25$; $x^2 + 16 = 25$; $x^2 = 9$. This equation also has two solutions: $x = 3$ or $x = -3$. Pairing up the y's and their respective x's, you get the four different solutions. The circle and parabola intersect in four distinct points.

6 Find the common solutions of $x^2 + y^2 = 9$ and $5x^2 - 6y = 18$. The answer is $\left(\dfrac{12}{5}, \dfrac{9}{5}\right), \left(-\dfrac{12}{5}, \dfrac{9}{5}\right)$, **(0, –3).**

Eliminate the x terms: Multiply the terms of the first equation by -5 (which gives you $-5x^2 - 5y^2 = -45$) and add the two equations together. The resulting equation is $-5y^2 - 6y = -27$. Rewrite the equation, setting it equal to 0, and factor. You get $0 = 5y^2 + 6y - 27 = (5y - 9)(y + 3)$. Use the multiplication property of zero to solve for the two solutions of this equation.

Replacing the y in the second equation (the equation of the parabola) with $\dfrac{9}{5}$, you get

$$5x^2 - 6\left(\frac{9}{5}\right) = 18$$

$$5x^2 - \frac{54}{5} = 18$$

$$5x^2 = \frac{144}{5}$$

Then divide both sides of the equation by 5 and take the square root of each side:

$$x^2 = \frac{144}{25}$$

$$x = \pm\frac{12}{5}$$

To find the other solution, let the y in the equation of the parabola be equal to -3. You get $5x^2 - 6(-3) = 18$; $5x^2 + 18 = 18$; $5x^2 = 0$. So, $x = 0$. The circle and parabola intersect or touch in three distinct points.

7 Find the common solutions of $x^2 + y^2 = 5$ and $y^2 = -4x$. The answer is **(–1, 2), (–1, –2).**

Replace the y^2 in the first equation with $-4x$. Rewrite the equation $x^2 - 4x = 5$ as $x^2 - 4x - 5 = 0$, which factors into $(x - 5)(x + 1) = 0$. The two solutions of this equation are $x = 5$ or $x = -1$. Replace the x with 5 in the second equation, and you get the equation $y^2 = -4(5) = -20$. But $y^2 = -20$ doesn't have any real solution. So, when $x = 5$, there isn't a common solution. You can use $x = 5$ in the equation of the circle, but you can't use $x = 5$ in the equation of the parabola. Now let $x = -1$, and you get $y^2 = -4(-1) = 4$; $y^2 = 4$. The two solutions of this equation are 2 and -2. The circle and parabola intersect only twice.

8 Find the common solutions of $x^2 + y^2 = 9$ and $y = x^2 + 11$. **No solution.**

Solve for x^2 in the second equation and replace the x^2 in the first equation with $y - 11$. You get $y - 11 + y^2 = 9$; $y^2 + y - 20 = 0$. Factor, and you find that $(y + 5)(y - 4) = 0$. The solutions are $y = -5$ or $y = 4$. When you replace the y in the parabola with -5, you get $-5 = x^2 + 11$; $-16 = x^2$. This equation has no real solution. You don't fare any better letting $y = 4$. You get $4 = x^2 + 11$; $x^2 = -7$. This entire system has no solution. The circle and parabola never cross or touch.

9 Find the common solution(s) of $y = 3^{x-1}$ and $y = 9^x$. The answer is $\left(-1, \dfrac{1}{9}\right)$.

Set y equal to y to get $3^{x-1} = 9^x$. Change the 9 to 3^2 and simplify: $3^{x-1} = (3^2)^x = 3^{2x}$. Now that the bases are the same, you can set the two exponents equal to one another and solve for x: $x - 1 = x$; $x = -1$. Replacing the x with -1 in $y = 3^{x-1}$, you get $y = 3^{-1-1} = 3^{-2} = \dfrac{1}{9}$.

10 Find the common solution(s) of $y = 8^{2-x}$ and $y = 4^{x^2-x}$. The answer is $\left(\frac{3}{2}, 2\sqrt{2}\right)$, **(-2, 4,096).**

First, substitute the exponential expression in the first equation for y in the second. Then change the 8 and 4 to powers of 2, and simplify the equation: $8^{2-x} = 4^{x^2-x}$, $(2^3)^{2-x} = (2^2)^{x^2-x}$, $2^{6-3x} = 2^{2x^2-2x}$. The bases are the same, so set the exponents equal to one another: $6 - 3x = 2x^2 - 2x$ becomes $0 = 2x^2 + x - 6$. Factoring, you get $0 = (2x - 3)(x + 2)$. When $x = \frac{3}{2}$, $y = 8^{2-3/2} = 8^{1/2} = \sqrt{8} = 2\sqrt{2}$, and when $x = -2$, $y = 8^{2-(-2)} = 8^4 = 4{,}096$.

11 Find the common solution(s) of $y = 2^x$ and $y = 1 - x$. The answer is **(0, 1).**

The first equation is an exponential that rises steadily as the x-values increase. The graph of the second equation is a line that falls steadily from left to right. They intersect at a single point. With some careful selections of points, you can quickly determine their single common solution, $(0, 1)$. Replacing the x with 0 in the exponential gives you $y = 2^0 = 1$. And replacing the x with 0 in the line gives you $y = 1 - 0 = 1$.

12 Find the common solution(s) of $y = e^{x^2}$ and $y = e$. The answer is **(1, e), (-1, e).**

The exponential function is positive for all values of x that you input. And the line is horizontal with a y-intercept of $(0, 1)$. If you replace the x in the exponential with 1, you get $y = e^1$ or $y = e$. The same thing occurs when you replace the x with -1; the square of -1 is also 1, so you get the same y-value.

13 Determine the solution of the system of inequalities: $x + y \leq 10$ and $2x + y > 18$.

Draw the line $x + y = 10$ (in slope-intercept form, $y = -x + 10$) as a solid line, and draw a dashed line for $2x + y = 18$ (or $y = -2x + 18$). The test point $(0, 0)$ works for the inequality $x + y \leq 10$, so you shade to the left of that line. The test point $(0, 0)$ doesn't satisfy the inequality $2x + y > 18$, so you don't want that point included in its solution; therefore, shade to the right of that line. The intersection of the two shaded areas moves downward between the intersection of the two lines.

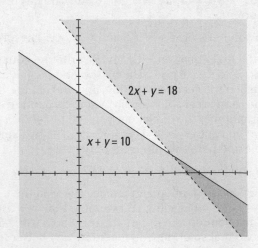

14 Determine the solution of the system of inequalities: $3x + 5y \leq 15$, $x \geq 0$, and $y \geq 0$.

The lines are all solid, because the inequalities all have an *or equal to* in them. The $x \geq 0$ and $y \geq 0$ indicate that all the points in the solution are in the first quadrant — where the x- and y-coordinates are positive. The test point $(0, 0)$ works for the inequality $3x + 5y \leq 15$, so the shaded area is below and to the left of the line $3x + 5y = 15$ (in slope-intercept form, $y = -\frac{3}{5}x + 3$).

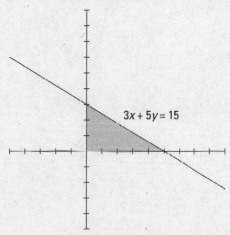

15 Determine the solution of the system of inequalities: $2x - y \leq 22$, $x \leq 2$, and $y \leq 0$.

The inequality $x \leq 2$ includes all the points to the left of the line $x = 2$, and the inequality $y \leq 0$ includes all the points below the x-axis. Rewrite the inequality $2x - y \leq 22$ in slope-intercept form by subtracting $2x$ from each side and then multiplying each side by -1. Don't forget to reverse the inequality. You get $y \geq 2x - 22$. The test point $(0, 0)$ works for $y \geq 2x - 22$ (you draw the solid line $y = 2x - 22$), so the test point is in the shaded area. The solutions for that inequality are to the left of and above that line. The points that meet all the qualifications are below the x-axis, left of $x = 2$, and above the diagonal line.

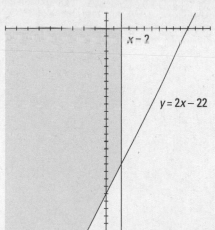

16 Determine the solution of the system of inequalities: $y \geq 2x^2 - 3x + 4$ and $x \geq y - 4$.

The inequality $y \geq 2x^2 - 3x + 4$ doesn't include the test point $(0, 0)$, so the points under the parabola don't work for the solution. All the points above or inside the parabola are part of the solution. The test point $(0, 0)$ does work for $x \geq y - 4$ (in slope-intercept form, the line is $y = x + 4$), so all the points below that line are in its solution. The solution to this system consists of the points beneath the line and inside the parabola.

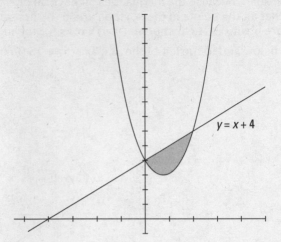

Part IV
Other Good Stuff: Lists, Arrays, and Imaginary Numbers

Sequence Type	Rule for Sequence Terms	Formula for Sum of Terms
Arithmetic	$a_n = a_1 + (n-1)d$	$S_n = \dfrac{n}{2}\left[2a_1 + (n-1)d\right] = \dfrac{n}{2}\left[a_1 + a_n\right]$
Geometric	$g_n = g_1(r)^{n-1}$	$S_n = \dfrac{g_1(1-r^n)}{1-r}$
Converging geometric	$g_n = g_1(r)^{n-1}, |r| < 1$	$S_\infty = \dfrac{g_1}{1-r}$
Odd numbers	$a_n = 2n-1$	$S_n = n^2$

Work with matrices in a free article at www.dummies.com/extras/algebraiiwb.

In this part . . .

✔ Move from the real to the imaginary with *i* and complex numbers.

✔ Solve equations with complex solutions and finish problems using the quadratic formula.

✔ Perform matrix operations.

✔ Use matrices to solve systems of linear equations.

✔ Work with patterns of numbers by writing sequences and their corresponding series.

✔ Count elements in sets using permutations and combinations.

Chapter 13

Getting More Complex with Imaginary Numbers

In This Chapter

▶ Keeping an eye on the imaginary i and its powers

▶ Operating on complex numbers: Addition, subtraction, and multiplication

▶ Complex division: Multiplying by conjugates

▶ Solving equations for complex solutions

The real numbers satisfied mathematicians and other scientists for centuries, but mathematicians eventually came across equations that couldn't be solved and problems that couldn't be answered without wandering into the realm of imaginary numbers.

Imaginary numbers involve the symbol i, which represents $\sqrt{-1}$. Put imaginary numbers together with real numbers, and you get *complex* numbers. The complex number system consists of numbers of the form $a + bi$, with a and b representing real numbers.

In this chapter, you investigate the powers of i, perform operations on complex numbers, write the answers in the standard form, and finally, solve equations such as $x^2 = -16$. Until the advent of imaginary numbers, equations such as this had no solution at all. Good thing they have solutions today, though, because some scientists and engineers use imaginary numbers on a regular basis to solve real world problems. And, hey, if you understand complex numbers, you may be well on your way to that dream career in quantum mechanics!

Simplifying Powers of i

Performing operations on complex numbers requires multiplying by i and simplifying powers of i. By definition, $i = \sqrt{-1}$, so $i^2 = -1$. If you want i^3, you compute it by writing $i^3 = i^2 \cdot i = -1 \cdot i = -i$. Also, $i^4 = i^2 \cdot i^2 = (-1)(-1) = 1$. And then the values of the powers start repeating themselves, because $i^5 = i$, $i^6 = -1$, $i^7 = -i$, and $i^8 = 1$. So, what do you do if you want a higher power, such as i^{345}, or something else pretty high up there? You don't want to have to write out all the powers up to i^{345} using the pattern (not when you could be white-water rafting or cleaning your room or watching the Cubs win the World Series!). Instead, use the following rule.

To compute the value of a power of i, determine whether the power is a multiple of 4, one more than a multiple of 4, two more than, or three more than a multiple of 4. Then apply the following:

✔ $i^{4n} = 1$

✔ $i^{4n+1} = i$

✔ $i^{4n+2} = -1$

✔ $i^{4n+3} = -i$

Q. Simplify the following: i^{444}, $i^{3,003}$, $i^{54,321}$, and $i^{111,002}$

A. $i^{444} = 1$; $i^{3,003} = -i$; $i^{54,321} = i$; $i^{111,002} = -1$. Writing the power of i as a multiple of 4 and what's left over (you know, the remainder), you get $i^{444} = i^{4(111)} = 1$, $i^{3,003} = i^{4(750)+3} = -i$, $i^{54,321} = i^{4(13,580)+1} = i$, and $i^{111,002} = i^{4(27,750)+2} = -1$.

1. Simplify: i^{45}

2. Simplify: i^{60}

3. Simplify: $i^{4,007}$

4. Simplify: $i^{2,002}$

Not Quite Brain Surgery: Doing Operations on Complex Numbers

A complex number has the standard form $a + bi$, where a and b are real numbers. You can add, subtract, and multiply complex numbers using the same algebraic rules as those for real numbers and then simplify the final answer so it's in the standard form.

For the most part, the i works just like any other variable (except you can simplify powers of i, as I mention in the preceding section). In general, you combine all real numbers, change all powers of i to 1, –1, i, or $-i$, and then combine all terms with i's in them. Division of complex numbers is something else, so division gets its own section — the next section in this chapter.

Q. Given the two complex numbers $3 + 4i$ and $-5 + 2i$, perform three separate operations: Add them, subtract the second from the first, and multiply them together

A. **Add: $-2 + 6i$; subtract: $8 + 2i$; multiply: $-23 - 14i$.** Adding, you have $(3 + 4i) + (-5 + 2i)$. Rearrange the terms so that the like terms are together. You get $3 - 5 + 4i + 2i = -2 + 6i$. Subtracting, you have to negate each term in the second complex number: $(3 + 4i) - (-5 + 2i) = 3 + 4i + 5 - 2i = 3 + 5 + 4i - 2i = 8 + 2i$. To multiply the complex numbers, first use the FOIL method (*First, Outer, Inner, Last* — see Chapter 1): $(3 + 4i)(-5 + 2i) = -15 + 6i - 20i + 8i^2 = -15 - 14i + 8i^2$. However, this answer isn't in standard form; you have to simplify the squared term. Remember that $i^2 = -1$, so write the answer as $-15 - 14i + 8(-1) = -15 - 14i - 8 = -23 - 14i$.

---

5. Compute: $(2 + 3i) + (-3 + 4i)$

6. Compute: $(-6 + i) - (3 - 2i)$

---

7. Compute: $(-8 + 6i) - 4(-1 + i)$

8. Compute: $(-3 + 2i)(4 - 9i)$

---

9. Compute: $(-6 + 3i)(6 + 3i)$

10. Compute: $(2 - 3i)^2$

"Dividing" Complex Numbers with a Conjugate

Mathematicians (that's you) can add, subtract, and multiply complex numbers. Technically, you can't divide complex numbers — in the traditional sense. You divide complex numbers by writing the division problem as a fraction and then multiplying the numerator and denominator by a conjugate.

The *conjugate* of the complex number $a + bi$ is $a - bi$.

The product of $(a + bi)(a - bi)$ is $a^2 + b^2$. How does that happen? Where's the i? Look at the steps in the multiplication: $(a + bi)(a - bi) = a^2 - abi + abi - b^2i^2 = a^2 - b^2(-1) = a^2 + b^2$, which is a real number — with no complex part. So when you need to divide one complex number by another, you multiply the numerator and denominator of the problem by the conjugate of the denominator. This step creates a real number in the denominator of the answer, which allows you to write the answer in the standard form of a complex number.

Q. Divide $10 + 5i$ by $4 - 3i$.

A. $1 + 2i$. Express the division as a fraction. Multiply both the numerator and denominator of the fraction by $4 + 3i$. You get $\frac{10+5i}{4-3i} \cdot \frac{4+3i}{4+3i} = \frac{40+50i+15i^2}{16+9} = \frac{40-15+50i}{25} = \frac{25+50i}{25}$.

Writing the answer in standard form, you get $1 + 2i$.

---

11. Divide $40 - 20i$ by $3 + i$.

12. Divide $5 + 10i$ by $2 - i$.

13. Divide $20 - 10i$ by $-3 + 4i$.

14. Divide $20i$ by $2 + 6i$.

Solving Equations with Complex Solutions

You often come across equations that have no real solutions — or equations that have the potential for many more real solutions than they actually have. For instance, the equation $x^2 + 1 = 0$ has no real solutions. If you write it as $x^2 = -1$ and try to take the square root of each side, you run into trouble. Not until you have the imaginary numbers can you write that the solution of this equation is $x = \pm i$. The equation has two complex solutions.

An example of an equation without enough real solutions is $x^4 - 81 = 0$. This equation factors into $(x^2 - 9)(x^2 + 9) = 0$. The two real solutions of this equation are 3 and –3. The two complex solutions are $3i$ and $-3i$.

To solve for the complex solutions of an equation, you use factoring, the square root property for solving quadratics, and the quadratic formula. Refer to Chapter 2 for a rundown of these techniques.

Q. Find all the roots, real and complex, of the equation $x^3 - 2x^2 + 25x - 50 = 0$.

A. $x = 2, 5i, -5i$. First, factor the equation to get $x^2(x - 2) + 25(x - 2) = (x - 2)(x^2 + 25) = 0$. Using the multiplication property of zero, you determine that $x - 2 = 0$ and $x = 2$. You also get $x^2 + 25 = 0$ and $x^2 = -25$. Take the square root of each side, and $x = \pm\sqrt{-25}$. Simplify the radical, using the equivalence for i, and the complex solutions are $\pm\sqrt{-25} = \pm\sqrt{25}\sqrt{-1} = \pm 5i$. The real root is 2, and the imaginary roots are $5i$ and $-5i$.

Q. Find all the roots, real and imaginary, of the equation $5x^2 - 8x + 5 = 0$.

A. $x = 0.4 + 0.6i, 0.4 - 0.6i$. The quadratic doesn't factor, so you use the quadratic formula:

$$x = \frac{-(-8) \pm \sqrt{(-8)^2 - 4(5)(5)}}{2(5)}$$

$$= \frac{8 \pm \sqrt{64 - 100}}{10}$$

$$= \frac{8 \pm \sqrt{-36}}{10}$$

$$= \frac{8 \pm 6i}{10} = \frac{4}{10} \pm \frac{6}{10}i$$

The only two solutions are complex: $0.4 + 0.6i$ and $0.4 - 0.6i$.

---

15. Find all the roots, real and imaginary, of $x^2 + 9 = 0$.

16. Find all the roots, real and imaginary, of $x^2 + 4x + 7 = 0$.

17. Find all the roots, real and imaginary, of $5x^2 + 6x + 3 = 0$.

18. Find all the roots, real and imaginary, of $x^4 + 12x^2 - 64 = 0$.

Answers to Problems on Imaginary Numbers

The following are the answers to the practice problems presented earlier in this chapter.

1 Simplify the following: i^{45}. The answer is ***i***.

Rewrite the term as $i^{4(11)+1} = i$.

2 Simplify the following: i^{60}. The answer is **1**.

Rewrite the term as $i^{4(15)} = 1$.

3 Simplify the following: $i^{4,007}$. The answer is ***–i***.

Rewrite the term as $i^{4(1,001)+3} = -i$.

4 Simplify the following: $i^{2,002}$. The answer is ***–1***.

Rewrite the term as $i^{4(5,00)+2} = -1$.

5 Compute: $(2 + 3i) + (-3 + 4i)$. The answer is ***–1 + 7i***.

Group the real parts and imaginary parts and simplify: $2 - 3 + 3i + 4i = -1 + 7i$.

6 Compute: $(-6 + i) - (3 - 2i)$. The answer is ***–9 + 3i***.

Distribute the negative sign over the second complex number; then group the real parts and imaginary parts and simplify: $-6 + i - 3 + 2i = -6 - 3 + i + 2i = -9 + 3i$.

7 Compute: $(-8 + 6i) - 4(-1 + i)$. The answer is ***–4 + 2i***.

Distribute the -4 over the second complex number; then group the real parts and imaginary parts and simplify: $-8 + 6i + 4 - 4i = -8 + 4 + 6i - 4i = -4 + 2i$.

8 Compute: $(-3 + 2i)(4 - 9i)$. The answer is **6 + 35*i***.

Multiply the two complex numbers together using FOIL. Replace the i^2 with -1. Then group the real parts and imaginary parts and simplify: $-12 + 27i + 8i - 18i^2 = -12 + 27i + 8i - 18(-1) = -12 + 18 + 27i + 8i = 6 + 35i$.

9 Compute: $(-6 + 3i)(6 + 3i)$. The answer is **–45**.

Multiply the two binomials together using FOIL. Combine the two opposite terms to get 0. Replace the i^2 with -1 and simplify: $36 - 18i + 18i + 9i^2 = -36 + 9(-1) = -36 - 9 = -45$.

10 Compute: $(2 - 3i)^2$. The answer is ***–5 – 12i***.

Multiply the two binomials together using FOIL. Replace the i^2 with -1. Then group the real parts and imaginary parts and simplify: $4 - 6i - 6i + 9i^2 = 4 - 6i - 6i + 9(-1) = 4 - 9 - 6i - 6i = -5 - 12i$.

11 Divide $40 - 20i$ by $3 + i$. The answer is **10 – 10*i***.

Write the problem as a fraction. Then multiply both the numerator and denominator by the conjugate of the denominator, $3 - i$. Simplify, and write the final answer in $a + bi$ form:

$$\frac{40-20i}{3+i} \cdot \frac{3-i}{3-i} = \frac{120-40i-60i+20i^2}{9+1} = \frac{120-20-100i}{10} = \frac{100-100i}{10} = 10-10i$$

12 Divide $5 + 10i$ by $2 - i$. The answer is **5*i***.

Write the problem as a fraction. Then multiply both the numerator and denominator by the conjugate of the denominator, $2 + i$. Simplify, and write the final answer in $a + bi$ form:

$$\frac{5+10i}{2-i} \cdot \frac{2+i}{2+i} = \frac{10+5i+20i+10i^2}{4+1} = \frac{10-10+25i}{5} = \frac{0+25i}{5} = 0+5i$$

13 Divide $20 - 10i$ by $-3 + 4i$. The answer is **$-4 - 2i$.**

Write the problem as a fraction. Then multiply both the numerator and denominator by the conjugate of the denominator, $-3 - 4i$. Simplify, and write the final answer in $a + bi$ form:

$$\frac{20-10i}{-3+4i} \cdot \frac{-3-4i}{-3-4i} = \frac{-60-80i+30i+40i^2}{9+16} = \frac{-60-40-50i}{25} = \frac{-100-50i}{25} = -4-2i$$

14 Divide $20i$ by $2 + 6i$. The answer is **$3 + i$.**

Write the problem as a fraction. Then multiply both the numerator and denominator by the conjugate of the denominator, $2 - 6i$. Simplify, and write the final answer in $a + bi$ form:

$$\frac{20i}{2+6i} \cdot \frac{2-6i}{2-6i} = \frac{40i-120i^2}{4+36} = \frac{120+40i}{40} = 3+i$$

15 Find all the roots, real and imaginary, of $x^2 + 9 = 0$. The answer is **$x = 3i, -3i$.**

Add -9 to each side to get $x^2 = -9$. Take the square root of each side. Then simplify the expression using i for the negative under the radical: $x = \pm\sqrt{-9} = \pm\sqrt{9}\sqrt{-1} = \pm 3i$.

16 Find all the roots, real and imaginary, of $x^2 + 4x + 7 = 0$. The answer is $x = -2 + \sqrt{3}i, x = -2 - \sqrt{3}i$.

Use the quadratic formula to solve for x. Simplify the expression using i for the negative under the radical:

$$x = \frac{-4 \pm \sqrt{4^2 - 4(1)(7)}}{2(1)}$$

$$= \frac{-4 \pm \sqrt{16 - 28}}{2}$$

$$= \frac{-4 \pm \sqrt{-12}}{2} = \frac{-4 \pm \sqrt{4}\sqrt{3}\sqrt{-1}}{2}$$

$$= \frac{-4 \pm 2\sqrt{3}i}{2} = -2 \pm \sqrt{3}i$$

17 Find all the roots, real and imaginary, of $5x^2 + 6x + 3 = 0$. The answer is

$x = -\dfrac{3}{5} + \dfrac{\sqrt{6}}{5}i, x = -\dfrac{3}{5} - \dfrac{\sqrt{6}}{5}i.$

Use the quadratic formula to solve for x. Simplify the expression using i for the negative under the radical:

$$x = \frac{-6 \pm \sqrt{6^2 - 4(5)(3)}}{2(5)}$$

$$= \frac{-6 \pm \sqrt{36 - 60}}{10}$$

$$= \frac{-6 \pm \sqrt{-24}}{10} = \frac{-6 \pm \sqrt{4}\sqrt{6}\sqrt{-1}}{10}$$

$$= \frac{-6 \pm 2\sqrt{6}i}{10} = \frac{-6}{10} \pm \frac{2\sqrt{6}i}{10} = -\frac{3}{5} \pm \frac{\sqrt{6}}{5}i$$

18 Find all the roots, real and imaginary, of $x^4 + 12x^2 - 64 = 0$. The answer is **$x = 2, -2, 4i, -4i$.**

Factor the left side: $(x^2 + 16)(x^2 - 4) = (x^2 + 16)(x - 2)(x + 2) = 0$. Obtain the two real roots by setting $x - 2$ and $x + 2$ equal to 0. When $x^2 + 16 = 0$, you find that $x^2 = -16$. Taking the square root of each side and using i for the -1 under the radical gives you the two imaginary roots.

Chapter 14

Getting Squared Away with Matrices

· ·

In This Chapter

▶ Enter the matrix: Understanding arrays of numbers

▶ Performing simple operations on matrices

▶ Multiplying one matrix by another

▶ Working with identity matrices (and a sweet shortcut) to find inverses

▶ Using matrix inverses to solve systems of equations

· ·

*I*n algebra, a matrix is a rectangular array of numbers, not a *revolution* or something *reloaded* such as the computer-generated universe inhabited by Keanu Reeves. An array of numbers is a quick, down-and-dirty way of expressing a lot of information without going into a lot of detail. In the real world, matrices are a way of organizing business elements or transportation capabilities into a format that you can operate on and analyze.

In this chapter, you add, subtract, multiply, and divide (well, sort of divide) matrices — when allowed. I cover the rules governing operations on matrices. People also use matrices to solve large systems of linear equations with ease and efficiency, so you get to practice that type of solving technique firsthand.

Describing Dimensions and Types of Matrices

Matrices are rectangular arrays of numbers, usually denoted as being matrices by having the numbers surrounded by a bracket. The numbers in a matrix are its *elements*. Matrices are classified in several ways, the most immediately apparent being their *dimension,* or size.

The dimension of a matrix is written $r \times c$, where r is the number of rows in the matrix and c is the number of columns. The dimension may be 3×2 or 4×7 or 1×5 or some other combination of numbers. Here are the types of matrices:

  ✔ **Column matrix:** A column matrix has one column and any number of rows.

  ✔ **Row matrix:** A row matrix has one row and any number of columns.

  ✔ **Zero matrix:** A zero matrix is quite predictable — all its elements are 0s.

  ✔ **Square matrix:** A square matrix is just what the name suggests: The number of rows and columns is the same.

  ✔ **Identity matrix:** An identity matrix is a square matrix in which the elements along the diagonal that runs from the upper-left corner to the lower-right corner are all 1s, with all the other elements being 0s. An identity matrix is one of the more important matrices because it shows up in finding inverse matrices and in solving systems of equations (which I cover later in this chapter).

You name matrices using capital letters to distinguish one matrix from another.

Q. Identify the dimension of each matrix and determine any special classifications:

$$A = \begin{bmatrix} 2 & -3 & 0 & 4 \\ -1 & 1 & 0 & 5 \end{bmatrix}, B = \begin{bmatrix} 1 & 0 \\ 0 & 1 \end{bmatrix}, C = \begin{bmatrix} -1 \\ 0 \\ 2 \end{bmatrix}$$

A. List the number of rows first and then the number of columns:

✔ Matrix A: 2×4; no other special classification

✔ Matrix B: 2×2; square matrix, identity matrix

✔ Matrix C: 3×1; column matrix

1. Identify the dimension of the following matrix and determine any special classifications.

$$\begin{bmatrix} 1 & 0 & 0 & 0 \\ 0 & 1 & 0 & 0 \\ 0 & 0 & 1 & 0 \\ 0 & 0 & 0 & 1 \end{bmatrix}$$

2. Identify the dimension of the following matrix and determine any special classifications.

$$\begin{bmatrix} 0 & 0 & 0 & 0 & 0 & 0 \end{bmatrix}$$

3. Identify the dimension of the following matrix and determine any special classifications.

$$\begin{bmatrix} 4 & 3 \\ 0 & -1 \\ -1 & 7 \\ 5 & 5 \\ -3 & 0 \end{bmatrix}$$

4. Identify the dimension of the following matrix and determine any special classifications.

$$\begin{bmatrix} 1 & 0 \\ 1 & 6 \end{bmatrix}$$

Adding, Subtracting, and Doing Scalar Multiplication on Matrices

As long as matrices have the exact same dimensions, you can add them to or subtract them from one another. The dimensions have to be the same, because each element has to pair up with its counterpart in the other matrix; you perform the operation on those elements.

Scalar multiplication of matrices means multiplying every element in a particular matrix by some number. Because only one matrix is involved in the operation, the dimension doesn't matter in scalar multiplication — any dimension works.

Q. Given the matrices $A = \begin{bmatrix} -3 & 2 \\ 0 & 4 \\ 1 & -5 \\ 6 & -1 \end{bmatrix}, B = \begin{bmatrix} -4 & -2 \\ 3 & 3 \\ 1 & -5 \\ 8 & 2 \end{bmatrix}$, perform the following operations: $A + B$, $2B$, and $3A - 4B$.

A. $A+B = \begin{bmatrix} -3 & 2 \\ 0 & 4 \\ 1 & -5 \\ 6 & -1 \end{bmatrix} + \begin{bmatrix} -4 & -2 \\ 3 & 3 \\ 1 & -5 \\ 8 & 2 \end{bmatrix} = \begin{bmatrix} -3+(-4) & 2+(-2) \\ 0+3 & 4+3 \\ 1+1 & -5+(-5) \\ 6+8 & -1+2 \end{bmatrix} = \begin{bmatrix} -7 & 0 \\ 3 & 7 \\ 2 & -10 \\ 14 & 1 \end{bmatrix}$

$2B = 2\begin{bmatrix} -4 & -2 \\ 3 & 3 \\ 1 & -5 \\ 8 & 2 \end{bmatrix} = \begin{bmatrix} 2(-4) & 2(-2) \\ 2(3) & 2(3) \\ 2(1) & 2(-5) \\ 2(8) & 2(2) \end{bmatrix} = \begin{bmatrix} -8 & -4 \\ 6 & 6 \\ 2 & -10 \\ 16 & 4 \end{bmatrix}$

$3A - 4B = 3\begin{bmatrix} -3 & 2 \\ 0 & 4 \\ 1 & -5 \\ 6 & -1 \end{bmatrix} - 4\begin{bmatrix} -4 & -2 \\ 3 & 3 \\ 1 & -5 \\ 8 & 2 \end{bmatrix}$

$= \begin{bmatrix} 3(-3) & 3(2) \\ 3(0) & 3(4) \\ 3(1) & 3(-5) \\ 3(6) & 3(-1) \end{bmatrix} + \begin{bmatrix} -4(-4) & -4(-2) \\ -4(3) & -4(3) \\ -4(1) & -4(-5) \\ -4(8) & -4(2) \end{bmatrix}$

$= \begin{bmatrix} -9 & 6 \\ 0 & 12 \\ 3 & -15 \\ 18 & -3 \end{bmatrix} + \begin{bmatrix} 16 & 8 \\ -12 & -12 \\ -4 & 20 \\ -32 & -8 \end{bmatrix} = \begin{bmatrix} 7 & 14 \\ -12 & 0 \\ -1 & 5 \\ -14 & -11 \end{bmatrix}$

5. If $A = \begin{bmatrix} 2 & 3 & 0 \\ 4 & 1 & 5 \end{bmatrix}$ and $B = \begin{bmatrix} -1 & -2 & 0 \\ 1 & 0 & 2 \end{bmatrix}$, find $A - B$.

6. If $A = \begin{bmatrix} -3 & 6 \\ 2 & 2 \end{bmatrix}$ and $B = \begin{bmatrix} -1 & -2 \\ -3 & 2 \end{bmatrix}$, find $4A + 2B$.

7. If $A = \begin{bmatrix} 4 & -2 & 3 \\ 1 & 6 & -5 \\ 4 & 4 & -2 \end{bmatrix}$ and $B = \begin{bmatrix} -4 & 2 & -3 \\ -1 & -6 & 5 \\ -4 & -4 & 2 \end{bmatrix}$, find $A + B$.

8. If $A = \begin{bmatrix} 3 & 0 & 0 \\ 0 & 0 & 7 \\ 0 & 5 & 1 \end{bmatrix}$ and $B = \begin{bmatrix} 1 & 0 & 0 \\ 0 & 1 & 0 \\ 0 & 0 & 1 \end{bmatrix}$, find $9A - B$

Trying Times: Multiplying Matrices by Each Other

You can multiply two matrices together only if the number of *columns* in the first matrix is equal to the number of *rows* in the second matrix. The resulting matrix has a dimension determined by the number of rows in the first matrix and the number of columns in the second matrix.

If matrix A has dimension $m \times n$ and matrix B has dimension $p \times q$, then the product $A \cdot B$ is possible only if $n = p$. Furthermore, the dimensions of the matrix formed from $A \cdot B$ (the answer matrix) are $m \times q$.

To multiply the matrices together, you sum up the products. The element in the first row, first column of the answer matrix is the result of multiplying each element in the first *row* of the first matrix times the corresponding elements in the first *column* of the second matrix — and finding the sum of all those products. You have to find n products (where n is the number of columns in the first matrix and the number of rows in the second) and add them together just to put a single number in the answer matrix.

One more example of how the numbers match up: The element in the third row, fourth column of the answer matrix is the result of multiplying all the elements in the third row of the first matrix times the fourth column of the second matrix — and summing up the products. Check out an example.

Q. Find the product $A \cdot B$ if $A = \begin{bmatrix} 2 & 0 \\ -1 & 3 \\ 0 & 1 \end{bmatrix}$ and $B = \begin{bmatrix} -2 & 1 \\ 1 & 0 \end{bmatrix}$.

A. **The resulting matrix is** $\begin{bmatrix} -4 & 2 \\ 5 & -1 \\ 1 & 0 \end{bmatrix}$. Matrix A is 3×2, and matrix B is 2×2. The number of columns in A matches the number of rows in B, so you can multiply the matrices together. The resulting matrix is 3×2. The following are the sums of the products you need to produce the elements in the answer matrix:

	Column 1	**Column 2**
Row 1	First row of A times first column of B $2(-2) + 0(1) = -4 + 0 = -4$	First row of A times second column of B $2(1) + 0(0) = 2 + 0 = 2$
Row 2	Second row of A times first column of B $-1(-2) + 3(1) = 2 + 3 = 5$	Second row of A times second column of B $-1(1) + 3(0) = -1 + 0 = -1$
Row 3	Third row of A times first column of B $0(-2) + 1(1) = 0 + 1 = 1$	Third row of A times second column of B $0(1) + 1(0) = 0 + 0 = 0$

9. Find the product of the two matrices

$A = \begin{bmatrix} 2 & 1 \\ 0 & 4 \end{bmatrix}$ and $B = \begin{bmatrix} 3 & -2 & 4 \\ 1 & 0 & 2 \end{bmatrix}$.

10. Find the product of the two matrices

$C = \begin{bmatrix} 1 & 0 \\ 4 & 3 \\ 2 & -2 \end{bmatrix}$ and $A = \begin{bmatrix} 2 & 1 \\ 0 & 4 \end{bmatrix}$.

11. Find the product BC:

$B = \begin{bmatrix} 3 & -2 & 4 \\ 1 & 0 & 2 \end{bmatrix}$ and $C = \begin{bmatrix} 1 & 0 \\ 4 & 3 \\ 2 & -2 \end{bmatrix}$

12. Find the product of the two matrices

$A = \begin{bmatrix} 2 & 1 \\ 0 & 4 \end{bmatrix}$ and $I = \begin{bmatrix} 1 & 0 \\ 0 & 1 \end{bmatrix}$.

The Search for Identity: Finding Inverse Matrices

The inverse of matrix A is matrix A^{-1}. Two matrices are *inverses* if their product is an identity matrix (see "Describing Dimensions and Types of Matrices," earlier in this chapter). Only square matrices have inverses, but not all square matrices have inverses. So, in terms of matrices, having an inverse is a pretty special characteristic.

The matrices $B = \begin{bmatrix} 4 & 5 \\ 5 & 6 \end{bmatrix}$ and $B^{-1} = \begin{bmatrix} -6 & 5 \\ 5 & -4 \end{bmatrix}$ are inverses of one another. In general, matrix

multiplication is *not* commutative — you can't change the order of multiplication and expect to get the same answer. However, with inverses, the order you multiply them in doesn't matter — you get the same result, and that result is an identity matrix:

$$B * B^{-1} = \begin{bmatrix} 4 & 5 \\ 5 & 6 \end{bmatrix} \cdot \begin{bmatrix} -6 & 5 \\ 5 & -4 \end{bmatrix} = \begin{bmatrix} -24+25 & 20-20 \\ -30+30 & 25-24 \end{bmatrix} = \begin{bmatrix} 1 & 0 \\ 0 & 1 \end{bmatrix}$$

$$B^{-1} * B = \begin{bmatrix} -6 & 5 \\ 5 & -4 \end{bmatrix} \cdot \begin{bmatrix} 4 & 5 \\ 5 & 6 \end{bmatrix} = \begin{bmatrix} -24+25 & -30+30 \\ 20-20 & 25-24 \end{bmatrix} = \begin{bmatrix} 1 & 0 \\ 0 & 1 \end{bmatrix}$$

If you have a matrix and you want to find its inverse (if it has one), you can use a method in which you put your matrix and an identity matrix of the same dimensions side by side in one big matrix. Then you change the elements in your original square matrix until they look like an identity matrix by performing operations such as multiplying and adding rows. The operations you can perform are exactly those you use when solving systems of equations: You can add two rows together to eliminate one of the numbers (when you add opposites), or you can multiply a row by some number and add it to another row. The result is the inverse of the original matrix sitting alongside the identity matrix. Here are some guidelines on making the transformation:

✔ Start with the first column on the left and work your way to the right; finish work on one column before moving to the next.

✔ When you start work on a new column, get the 1 in its proper place before dealing with the 0s (1s should start in the upper left and go down along a diagonal). Figure out what number you can multiply an element by to get a product of 1 (find the element's reciprocal) and then multiply the entire row by that number.

✔ Change numbers to 0s using the elimination process (see Chapter 11). Basically, figure out what number you'd have to add to an element to get a sum of 0. Multiply one of the other rows by that number and add the two rows together.

If you're working on the first column, multiply row one by that magic elimination number; the second column, multiply row two; third column, row three; and so on.

In the examples, I show you the step-by-step process for finding the inverse of a square matrix (plus a slick shortcut for 2×2 matrices).

Q. Find the inverse of the following matrix:

$$A = \begin{bmatrix} 1 & 0 & 4 \\ -2 & 1 & -6 \\ 3 & 1 & 9 \end{bmatrix}$$

A. Use the following steps.

1. Write the original matrix and an identity matrix all in one array:

$$\begin{bmatrix} 1 & 0 & 4 & \vdots & 1 & 0 & 0 \\ -2 & 1 & -6 & \vdots & 0 & 1 & 0 \\ 3 & 1 & 9 & \vdots & 0 & 0 & 1 \end{bmatrix}$$

2. Change the first column of the original matrix so it looks like that of an identity matrix.

You don't have to do anything to the first row, because you want that 1 in the upper-left corner. However, you do need the other two elements in the first column to be 0s. The second element is –2, so to get a sum of 0, you have to add 2 to it. Add 2 times row one to row two:

$$\begin{bmatrix} 1 & 0 & 4 & \vdots & 1 & 0 & 0 \\ 0 & 1 & 2 & \vdots & 2 & 1 & 0 \\ 3 & 1 & 9 & \vdots & 0 & 0 & 1 \end{bmatrix}$$

The third element in this column is 3. To change it to 0, multiply each element in row one by –3 and add the result to row three; this sum becomes your new row three:

$$\begin{bmatrix} 1 & 0 & 4 & \vdots & 1 & 0 & 0 \\ 0 & 1 & 2 & \vdots & 2 & 1 & 0 \\ 0 & 1 & -3 & \vdots & -3 & 0 & 1 \end{bmatrix}$$

3. Make the matrix's second column look like the second column of an identity matrix.

   First, you need a 1 in the second row, second column. Luckily, your other calculations already put a 1 there. You already have a 0 above it, so you only need to work on putting a 0 in the third row. Add –1 times row two to row three, and write the sum as the new row three:

$$\begin{bmatrix} 1 & 0 & 4 & : & 1 & 0 & 0 \\ 0 & 1 & 2 & : & 2 & 1 & 0 \\ 0 & 0 & -5 & : & -5 & -1 & 1 \end{bmatrix}$$

4. Make the third column of the matrix look like that of an identity matrix.

   You need a 1 in the third row, third column. This element is currently –5, so you need to divide every element in that row by –5 (multiply by $-\frac{1}{5}$):

$$\begin{bmatrix} 1 & 0 & 4 & : & 1 & 0 & 0 \\ 0 & 1 & 2 & : & 2 & 1 & 0 \\ 0 & 1 & 1 & : & 1 & 0.2 & -0.2 \end{bmatrix}$$

   The other terms need to be 0s, so use the elimination method: Add –4 times row three to row one:

$$\begin{bmatrix} 1 & 0 & 0 & : & -3 & -0.8 & 0.8 \\ 0 & 1 & 2 & : & 2 & 1 & 0 \\ 0 & 0 & 1 & : & 1 & 0.2 & -0.2 \end{bmatrix}$$

   Add –2 times row three to row two:

$$\begin{bmatrix} 1 & 0 & 0 & : & -3 & -0.8 & 0.8 \\ 0 & 1 & 0 & : & 0 & 0.6 & 0.4 \\ 0 & 0 & 1 & : & 1 & 0.2 & -0.2 \end{bmatrix}$$

5. You now have an identity matrix to the left. The matrix on the right is the inverse of the original matrix:

$$A^{-1} = \begin{bmatrix} -3 & -0.8 & 0.8 \\ 0 & 0.6 & 0.4 \\ 1 & 0.2 & -0.2 \end{bmatrix}$$

Q. Find the inverse of the following matrix:

$$\begin{bmatrix} 5 & 8 \\ 7 & 11 \end{bmatrix}$$

A. The process I outline in the preceding example works for any size square matrix. But you can use a neater, quicker way to find the inverse of a 2 × 2 matrix.

The inverse of matrix $\begin{bmatrix} a & b \\ c & d \end{bmatrix}$ is $\begin{bmatrix} \frac{d}{m} & -\frac{b}{m} \\ -\frac{c}{m} & \frac{a}{m} \end{bmatrix}$, where $m = ad - bc$, the difference of the cross-products. Of course, if the difference is 0, the matrix has no inverse. Basically, you just swap the elements a and d, negate the elements b and c, and divide each element by the difference of the cross-products.

For this particular matrix, find the difference of the cross-products: $m = 5(11) - 8(7) = 55 - 56 = -1$. Now switch the 5 and 11 and negate both the 8 and the 7. Then divide each element by -1:

$$\begin{bmatrix} 11 & -8 \\ -7 & 5 \end{bmatrix} \rightarrow \begin{bmatrix} \frac{11}{-1} & -\frac{8}{-1} \\ -\frac{7}{-1} & \frac{5}{-1} \end{bmatrix} \rightarrow \begin{bmatrix} -11 & 8 \\ 7 & -5 \end{bmatrix}$$

The inverse of the original matrix is $\begin{bmatrix} -11 & 8 \\ 7 & -5 \end{bmatrix}$.

---

13. Find the inverse of the following matrix:

$$\begin{bmatrix} 3 & 2 \\ 7 & 5 \end{bmatrix}$$

14. Find the inverse of the following matrix

$$\begin{bmatrix} -4 & -5 \\ 7 & 9 \end{bmatrix}$$

15. Find the inverse of the following matrix:

$$\begin{bmatrix} 6 & 3 \\ 4 & 2 \end{bmatrix}$$

16. Find the inverse of the following matrix:

$$\begin{bmatrix} -8 & 4 \\ 5 & -2 \end{bmatrix}$$

17. Find the inverse of the following matrix:

$$\begin{bmatrix} 1 & 2 & 3 \\ 0 & 2 & 4 \\ -1 & 4 & -1 \end{bmatrix}$$

18. Find the inverse of the following matrix:

$$\begin{bmatrix} 1 & 2 & -3 \\ 3 & 5 & -7 \\ -1 & -3 & 6 \end{bmatrix}$$

Using Matrices to Solve Systems of Equations

One very nice application of matrices is using them to solve systems of linear equations. This application is especially helpful when you have large systems with lots of variables, as well as when you have a graphing calculator or computer program such as Excel to do all the computations for you.

When solving systems of equations using matrices, the system has to have the same number of equations as it has variables. To solve systems of linear equations using matrices, do the following:

1. **Write all the equations so the variables are in the same order, and set the equations equal to the constants.**

2. **Create a coefficient matrix made up of all the coefficients of the variables and a constant matrix made up of all the constants.**

3. **Find the solution of the system by multiplying the *inverse* of the coefficient matrix times the constant matrix.**

Here's an example to show the process in more detail.

Q. Solve the system of equations using matrices:

$$x - y - z = 7$$
$$-x + 2y - 3z = -7$$
$$x - 2y + z = 9$$

A. (2, –4, –1); or $x = 2$, $y = -4$, $z = -1$. First, create the coefficient matrix:

$$\begin{bmatrix} 1 & -1 & -1 \\ -1 & 2 & -3 \\ 1 & -2 & 1 \end{bmatrix}$$

Then find the inverse of that matrix using the techniques from the preceding section. The inverse of the coefficient matrix is

$$\begin{bmatrix} 2 & -1.5 & -2.5 \\ 1 & -1 & -2 \\ 0 & -0.5 & -0.5 \end{bmatrix}$$

Now create the *constant* matrix:

$$\begin{bmatrix} 7 \\ -7 \\ 9 \end{bmatrix}$$

Multiply the coefficient matrix times the constant matrix:

$$\begin{bmatrix} 2 & -1.5 & -2.5 \\ 1 & -1 & -2 \\ 0 & -0.5 & -0.5 \end{bmatrix} \cdot \begin{bmatrix} 7 \\ -7 \\ 9 \end{bmatrix} = \begin{bmatrix} 2 \\ -4 \\ -1 \end{bmatrix}$$

(Refer to "Trying Times: Multiplying Matrices by Each Other," earlier in the chapter, if you need help.) You see the solutions in the resulting matrix, going down, in the order in which you wrote the coefficients.

19. Solve the system of equations using matrices: $3x - 2y = 17$ and $-4x + 3y = -23$

20. Solve the system of equations using matrices: $-6x + 5y = 8$ and $-4x + 3y = 6$

21. Solve the system of equations using matrices: $x - y - z = -1$, $2x + 3y + z = 0$, and $3y + 2z = 2$

22. Solve the system of equations using matrices: $2x + y - z = 9$, $x + 2y - z = 6$, and $x - 2y + z = 0$

Answers to Problems on Matrices

The following are the answers to the practice problems presented earlier in this chapter.

1 Identify the dimension of the following matrix and determine any special classifications:

$$\begin{bmatrix} 1 & 0 & 0 & 0 \\ 0 & 1 & 0 & 0 \\ 0 & 0 & 1 & 0 \\ 0 & 0 & 0 & 1 \end{bmatrix}$$

The answer is **4 × 4; square, identity.**

The elements in the diagonal running from left to right are all 1s, and the rest of the elements are 0s, making this array an identity matrix.

2 Identify the dimension of the following matrix and determine any special classifications:

$$\begin{bmatrix} 0 & 0 & 0 & 0 & 0 & 0 \end{bmatrix}$$

The answer is **1 × 6; row, zero.**

This matrix has only one row, and all the elements are 0s.

3 Identify the dimension of the following matrix and determine any special classifications:

$$\begin{bmatrix} 4 & 3 \\ 0 & -1 \\ -1 & 7 \\ 5 & 5 \\ -3 & 0 \end{bmatrix}$$

The answer is **5 × 2; no special characteristics.**

This matrix has five rows and two columns.

4 Identify the dimension of the following matrix and determine any special classifications:

$$\begin{bmatrix} 1 & 0 \\ 1 & 6 \end{bmatrix}$$

The answer is **2 × 2; square.**

The number of rows and columns is the same.

5 If $A = \begin{bmatrix} 2 & 3 & 0 \\ 4 & 1 & 5 \end{bmatrix}$ and $B = \begin{bmatrix} -1 & -2 & 0 \\ 1 & 0 & 2 \end{bmatrix}$, find $A - B$. The answer is $\begin{bmatrix} \mathbf{3} & \mathbf{5} & \mathbf{0} \\ \mathbf{3} & \mathbf{1} & \mathbf{3} \end{bmatrix}$.

Subtract each element in the second matrix from the corresponding elements in the first matrix:

$$\begin{bmatrix} 2-(-1) & 3-(-2) & 0-0 \\ 4-1 & 1-0 & 5-2 \end{bmatrix} = \begin{bmatrix} 3 & 5 & 0 \\ 3 & 1 & 3 \end{bmatrix}$$

6 If $A = \begin{bmatrix} -3 & 6 \\ 2 & 2 \end{bmatrix}$ and $B = \begin{bmatrix} -1 & -2 \\ -3 & 2 \end{bmatrix}$, find $4A + 2B$. The answer is $\begin{bmatrix} -14 & 20 \\ 2 & 12 \end{bmatrix}$.

Multiply each element in the first matrix by 4 and all those in the second matrix by 2. Then add the respective elements together:

$$4 \begin{bmatrix} -3 & 6 \\ 2 & 2 \end{bmatrix} + 2 \begin{bmatrix} -1 & -2 \\ -3 & 2 \end{bmatrix} = \begin{bmatrix} -12+(-2) & 24+(-4) \\ 8+(-6) & 8+4 \end{bmatrix} = \begin{bmatrix} -14 & 20 \\ 2 & 12 \end{bmatrix}$$

7 If $A = \begin{bmatrix} 4 & -2 & 3 \\ 1 & 6 & -5 \\ 4 & 4 & -2 \end{bmatrix}$ and $B = \begin{bmatrix} -4 & 2 & -3 \\ -1 & -6 & 5 \\ -4 & -4 & 2 \end{bmatrix}$, find $A + B$. The answer is $\begin{bmatrix} 0 & 0 & 0 \\ 0 & 0 & 0 \\ 0 & 0 & 0 \end{bmatrix}$.

Add all the elements in the first matrix to their counterparts in the second matrix. Because each element is the opposite of what you add it to, the result is a zero matrix:

$$\begin{bmatrix} 4+(-4) & -2+2 & 3+(-3) \\ 1+(-1) & 6+(-6) & -5+5 \\ 4+(-4) & 4+(-4) & -2+2 \end{bmatrix} = \begin{bmatrix} 0 & 0 & 0 \\ 0 & 0 & 0 \\ 0 & 0 & 0 \end{bmatrix}$$

8 If $A = \begin{bmatrix} 3 & 0 & 0 \\ 0 & 0 & 7 \\ 0 & 5 & 1 \end{bmatrix}$ and $B = \begin{bmatrix} 1 & 0 & 0 \\ 0 & 1 & 0 \\ 0 & 0 & 1 \end{bmatrix}$, find $9A - B$. The answer is $\begin{bmatrix} 26 & 0 & 0 \\ 0 & -1 & 63 \\ 0 & 45 & 8 \end{bmatrix}$.

Multiply each element in the first matrix by 9. Then subtract all the elements in the second matrix from those results:

$$9 \begin{bmatrix} 3 & 0 & 0 \\ 0 & 0 & 7 \\ 0 & 5 & 1 \end{bmatrix} - \begin{bmatrix} 1 & 0 & 0 \\ 0 & 1 & 0 \\ 0 & 0 & 1 \end{bmatrix} = \begin{bmatrix} 27-1 & 0-0 & 0-0 \\ 0-0 & 0-1 & 63-0 \\ 0-0 & 45-0 & 9-1 \end{bmatrix} = \begin{bmatrix} 26 & 0 & 0 \\ 0 & -1 & 63 \\ 0 & 45 & 8 \end{bmatrix}$$

9 Find the product of the two matrices $A = \begin{bmatrix} 2 & 1 \\ 0 & 4 \end{bmatrix}$ and $B = \begin{bmatrix} 3 & -2 & 4 \\ 1 & 0 & 2 \end{bmatrix}$. The answer is $\begin{bmatrix} 7 & -4 & 10 \\ 4 & 0 & 8 \end{bmatrix}$.

The product of a 2×2 matrix times a 2×3 matrix is a 2×3 matrix. Find the sums of all the row-to-column products:

$$\begin{bmatrix} 2(3)+1(1) & 2(-2)+1(0) & 2(4)+1(2) \\ 0(3)+4(1) & 0(-2)+4(0) & 0(4)+4(2) \end{bmatrix} = \begin{bmatrix} 7 & -4 & 10 \\ 4 & 0 & 8 \end{bmatrix}$$

10 Find the product of the two matrices $C = \begin{bmatrix} 1 & 0 \\ 4 & 3 \\ 2 & -2 \end{bmatrix}$ and $A = \begin{bmatrix} 2 & 1 \\ 0 & 4 \end{bmatrix}$. The answer is $\begin{bmatrix} 2 & 1 \\ 8 & 16 \\ 4 & -6 \end{bmatrix}$.

The product of a 3×2 matrix times a 2×2 matrix is a 3×2 matrix. Find the sums of all the row-to-column products:

$$\begin{bmatrix} 1\,(2)+0\,(0) & 1\,(1)+0\,(4) \\ 4\,(2)+3\,(0) & 4\,(1)+3\,(4) \\ 2\,(2)+(-2)\,(0) & 2\,(1)+(-2)\,(4) \end{bmatrix} = \begin{bmatrix} 2 & 1 \\ 8 & 16 \\ 4 & -6 \end{bmatrix}$$

11 Find the product BC if $B = \begin{bmatrix} 3 & -2 & 4 \\ 1 & 0 & 2 \end{bmatrix}$ and $C = \begin{bmatrix} 1 & 0 \\ 4 & 3 \\ 2 & -2 \end{bmatrix}$. The answer is $\begin{bmatrix} 3 & -14 \\ 5 & -4 \end{bmatrix}$.

The product of a 2×3 matrix times a 3×2 matrix is a 2×2 matrix. Find the sums of all the row-to-column products:

$$\begin{bmatrix} 3(1)+(-2)(4)+4(2) & 3(0)+(-2)(3)+4(-2) \\ 1(1)+0(4)+2(2) & 1(0)+0(3)+2(-2) \end{bmatrix} = \begin{bmatrix} 3 & -14 \\ 5 & -4 \end{bmatrix}$$

12 Find the product of the two matrices $A = \begin{bmatrix} 2 & 1 \\ 0 & 4 \end{bmatrix}$ and $I = \begin{bmatrix} 1 & 0 \\ 0 & 1 \end{bmatrix}$. The answer is $\begin{bmatrix} 2 & 1 \\ 0 & 4 \end{bmatrix}$.

The product of a 2×2 matrix times a 2×2 matrix is a 2×2 matrix. Find the sums of all the row-to-column products. The second matrix is an identity matrix, which is why the result is actually the first, original matrix:

$$\begin{bmatrix} 2(1)+1(0) & 2(0)+1(1) \\ 0(1)+4(0) & 0(0)+4(1) \end{bmatrix} = \begin{bmatrix} 2 & 1 \\ 0 & 4 \end{bmatrix}$$

13 Find the inverse of the matrix $\begin{bmatrix} 3 & 2 \\ 7 & 5 \end{bmatrix}$. The answer is $\begin{bmatrix} 5 & -2 \\ -7 & 3 \end{bmatrix}$.

The difference between the cross-products, m, is $15 - 14 = 1$. Reverse the 3 and 5, and negate the 2 and 7. Because you're dividing by 1, the new entries don't change.

14 Find the inverse of the matrix $\begin{bmatrix} -4 & -5 \\ 7 & 9 \end{bmatrix}$. The answer is $\begin{bmatrix} -9 & -5 \\ 7 & 4 \end{bmatrix}$.

The difference between the cross-products, m, is $-36 - (-35) = -1$. Reverse the -4 and 9. Then negate the -5 to make it 5 and the 7 to make it -7. Dividing every term by -1, all the signs change — even the two that you changed already (they change back).

15 Find the inverse of the matrix $\begin{bmatrix} 6 & 3 \\ 4 & 2 \end{bmatrix}$. **No inverse.**

The difference between the cross-products, m, is $12 - 12 = 0$. This result means that this matrix has no inverse.

16 Find the inverse of the matrix $\begin{bmatrix} -8 & 4 \\ 5 & -2 \end{bmatrix}$. The answer is $\begin{bmatrix} \frac{1}{2} & 1 \\ \frac{5}{4} & 2 \end{bmatrix}$.

The difference between the cross-products, m, is $16 - 20 = -4$. Reverse the -8 and -2. Then change the 4 and 5 to negative numbers. Finally, divide each number by -4 and simplify.

17 Find the inverse of the matrix $\begin{bmatrix} 1 & 2 & 3 \\ 0 & 2 & 4 \\ -1 & 4 & -1 \end{bmatrix}$. The answer is $\begin{bmatrix} 9/10 & -7/10 & -1/10 \\ 1/5 & -1/10 & 1/5 \\ -1/10 & 3/10 & -1/10 \end{bmatrix}$.

Write the matrix and a 3×3 identity matrix next to it to make one big array.

Change the first column so it looks like an identity column: The element in the first row is already a 1 and the element in the second row is already a 0, so you just have to change the -1 to a 0. Add row one to row three :

$$\begin{bmatrix} 1 & 2 & 3 & : & 1 & 0 & 0 \\ 0 & 2 & 4 & : & 0 & 1 & 0 \\ -1 & 4 & -1 & : & 0 & 0 & 1 \end{bmatrix} \rightarrow \begin{bmatrix} 1 & 2 & 3 & : & 1 & 0 & 0 \\ 0 & 2 & 4 & : & 0 & 1 & 0 \\ 0 & 6 & 2 & : & 1 & 0 & 1 \end{bmatrix}$$

Change the second column so it looks like an identity column: The element in the second row, second column needs to be a 1, but it's currently a 2. Divide row two by 2. You need 0s above and below the 1, so you then multiply row two by -2 and add that answer to row one. Multiply row two by -6 and add it to row three:

$$\begin{bmatrix} 1 & 2 & 3 & : & 1 & 0 & 0 \\ 0 & 1 & 2 & : & 0 & 1/2 & 0 \\ 0 & 6 & 2 & : & 1 & 0 & 1 \end{bmatrix} \rightarrow \begin{bmatrix} 1 & 0 & -1 & : & 1 & -1 & 0 \\ 0 & 1 & 2 & : & 0 & 1/2 & 0 \\ 0 & 0 & -10 & : & 1 & -3 & 1 \end{bmatrix}$$

Change the third column so it looks like an identity column: The -10 in the third row, third column is changed to a 1 by dividing row three by -10. Then get 0s above the 1 by adding row three to row one. Multiply row three by -2 and add it to row two:

$$\begin{bmatrix} 1 & 0 & -1 & : & 1 & -1 & 0 \\ 0 & 1 & 2 & : & 0 & 1/2 & 0 \\ 0 & 0 & 1 & : & -1/10 & 3/10 & -1/10 \end{bmatrix} \rightarrow \begin{bmatrix} 1 & 0 & 0 & : & 9/10 & -7/10 & -1/10 \\ 0 & 1 & 0 & : & 1/5 & -1/10 & 1/5 \\ 0 & 0 & 1 & : & -1/10 & 3/10 & -1/10 \end{bmatrix}$$

18 Find the inverse of the matrix $\begin{bmatrix} 1 & 2 & -3 \\ 3 & 5 & -7 \\ -1 & -3 & 6 \end{bmatrix}$. The answer is $\begin{bmatrix} -9 & 3 & -1 \\ 11 & -3 & 2 \\ 4 & -1 & 1 \end{bmatrix}$.

Write the matrix and a 3×3 identity matrix next to it.

The first row, first column already has an element of 1, so you just have to get 0s below that 1. Multiply row one by -3 and add it to row two. Add row one to row three:

$$\begin{bmatrix} 1 & 2 & -3 & : & 1 & 0 & 0 \\ 3 & 5 & -7 & : & 0 & 1 & 0 \\ -1 & -3 & 6 & : & 0 & 0 & 1 \end{bmatrix} \rightarrow \begin{bmatrix} 1 & 2 & -3 & : & 1 & 0 & 0 \\ 0 & -1 & 2 & : & -3 & 1 & 0 \\ 0 & -1 & 3 & : & 1 & 0 & 1 \end{bmatrix}$$

Change the second column so it looks like an identity column: The element along the diagonal is currently a –1, so multiply row two by –1. You then want 0s above and below the 1 in the second row, second column, so multiply row two by –2 and add it to row one. Then add row two to row three:

$$\left[\begin{array}{ccc:ccc} 1 & 2 & -3 & 1 & 0 & 0 \\ 0 & 1 & -2 & 3 & -1 & 0 \\ 0 & -1 & 3 & 1 & 0 & 1 \end{array}\right] \rightarrow \left[\begin{array}{ccc:ccc} 1 & 0 & 1 & -5 & 2 & 0 \\ 0 & 1 & -2 & 3 & -1 & 0 \\ 0 & 0 & 1 & 4 & -1 & 1 \end{array}\right]$$

The third row, third column has a 1 for an element (after all the work on the first two columns). To get 0s above that 1, multiply row three by –1 and add it to row one. Then multiply row three by 2 and add it to row two:

$$\left[\begin{array}{ccc:ccc} 1 & 0 & 0 & -9 & 3 & -1 \\ 0 & 1 & 0 & 11 & -3 & 2 \\ 0 & 0 & 1 & 4 & -1 & 1 \end{array}\right]$$

19 Solve the system of equations using matrices: $3x - 2y = 17$ and $-4x + 3y = -23$. The answer is $x = 5, y = -1.$

Write the coefficient matrix and find its inverse. (Refer to the section on "The Search for Identity: Finding Inverse Matrices" for info on that technique.) Then multiply that inverse by the constant matrix. (You can find how to multiply matrices in the section "Trying Times: Multiplying Matrices by Each Other.") Here's how this problem plays out:

$$\begin{bmatrix} 3 & -2 \\ -4 & 3 \end{bmatrix}^{-1} = \begin{bmatrix} 3 & 2 \\ 4 & 3 \end{bmatrix}$$

$$\begin{bmatrix} 3 & 2 \\ 4 & 3 \end{bmatrix} \cdot \begin{bmatrix} 17 \\ -23 \end{bmatrix} = \begin{bmatrix} 5 \\ -1 \end{bmatrix}$$

20 Solve the system of equations using matrices: $-6x + 5y = 8$ and $-4x + 3y = 6$. The answer is $x = -3, y = -2.$

Write the coefficient matrix and find its inverse. Then multiply that inverse by the constant matrix:

$$\begin{bmatrix} -6 & 5 \\ -4 & 3 \end{bmatrix}^{-1} = \begin{bmatrix} 3/2 & -5/2 \\ 2 & -3 \end{bmatrix}$$

$$\begin{bmatrix} 3/2 & -5/2 \\ 2 & -3 \end{bmatrix} \cdot \begin{bmatrix} 8 \\ 6 \end{bmatrix} = \begin{bmatrix} -3 \\ -2 \end{bmatrix}$$

21 Solve the system of equations using matrices: $x - y - z = -1$, $2x + 3y + z = 0$, and $3y + 2z = 2$. The answer is $x = 1, y = -2, z = 4.$

Write the coefficient matrix and find its inverse. Then multiply that inverse by the constant matrix:

$$\begin{bmatrix} 1 & -1 & -1 \\ 2 & 3 & 1 \\ 0 & 3 & 2 \end{bmatrix}^{-1} = \begin{bmatrix} 3 & -1 & 2 \\ -4 & 2 & -3 \\ 6 & -3 & 5 \end{bmatrix}$$

$$\begin{bmatrix} 3 & -1 & 2 \\ -4 & 2 & -3 \\ 6 & -3 & 5 \end{bmatrix} \cdot \begin{bmatrix} -1 \\ 0 \\ 2 \end{bmatrix} = \begin{bmatrix} 1 \\ -2 \\ 4 \end{bmatrix}$$

22 Solve the system of equations using matrices: $2x + y - z = 9$, $x + 2y - z = 6$, and $x - 2y + z = 0$. The answer is $x = 3$, $y = 0$, $z = -3$.

Write the coefficient matrix and find its inverse. Then multiply that inverse by the constant matrix:

$$\begin{bmatrix} 2 & 1 & -1 \\ 1 & 2 & -1 \\ 1 & -2 & 1 \end{bmatrix}^{-1} = \begin{bmatrix} 0 & 0.5 & 0.5 \\ -1 & 1.5 & 0.5 \\ -2 & 2.5 & 1.5 \end{bmatrix}$$

$$\begin{bmatrix} 0 & 0.5 & 0.5 \\ -1 & 1.5 & 0.5 \\ -2 & 2.5 & 1.5 \end{bmatrix} \cdot \begin{bmatrix} 9 \\ 6 \\ 0 \end{bmatrix} = \begin{bmatrix} 3 \\ 0 \\ -3 \end{bmatrix}$$

Chapter 15

Going Out of Sequence with Sequences and Series

A *sequence* is a list of items all separated by commas (milk, sugar, eggs . . . okay, so it's generally a list of numbers). Mathematical sequences usually have a rule determining what number comes in which position in the list. A *series* is the sum of the numbers in a sequence — or that list. You add up a few or many or all the terms in a sequence.

In this chapter, you see arithmetic and geometric sequences — some of the more commonly used sequences. You also form many other types of sequences by just writing a rule using various mathematical operations. Finally, you figure out how to add the terms in a sequence using rules that apply to series.

Writing the Terms of a Sequence

The sequence 1, 3, 5, 7, 9, . . . consists of the odd positive integers. Notice that I stop listing the numbers at 9, hoping that you can determine the pattern from those few terms. The three dots, called an *ellipsis,* mean "and so on and so on." You often see the terms of a sequence written in *braces:* {1, 3, 5, 7, 9, . . . }; this mathematical notation is standard for sequences. Even graphing calculators recognize the braces as designating the list of numbers in a sequence.

You can also describe sequences using a rule. You let n represent a counting number and write $\{2n - 1\}$. This rule gives you all the terms in a sequence when you replace the n with each of the counting numbers, one after the other.

Q. Write the first six terms of the sequence: $\{n^3 - 2n + 1\}$

A. **0, 5, 22, 57, 116, and 205.** When you replace n with 1, you get $1^3 - 2(1) + 1 = 1 - 2 + 1 = 0$. When you replace n with 2, you get $2^3 - 2(2) + 1 = 8 - 4 + 1 = 5$. And so on — or should I use...?

Q. Find a rule that describes how to find all the terms in the sequence: 2, –6, 10, –14, 18, –22, . . .

A. $\{(-1)^{n+1}(4n - 2)\}$. You first notice that the terms alternate in sign between positive and negative. If you disregard the signs, the numbers by themselves (their absolute values) go up by 4 units each time. You deal with these two features separately. First, to make numbers go up by 4 each time, multiply an n by 4, giving you $4n$. But when you let $n = 1$, you get 4, and when it's 2, you get 8. These numbers are too big by 2, so subtract 2, and you get $4n - 2$. That takes care of the numerical values: $\{4n - 2\} = \{2, 6, 10, 14, 18, 22, . . .\}$. You deal with the alternating part by multiplying the numbers by powers of –1. When –1 is raised to an odd power, it's equal to –1. When you raise –1 to an even power, the number is equal to +1. When $n = 1$, you want a positive number in this sequence, so use $(-1)^{n+1}$ as a multiplier; write the sequence rule as $\{(-1)^{n+1}(4n - 2)\}$. As n changes, the power on the –1 alternates between even and odd, giving you the + and – signs for the terms.

1. Write the first six terms of the sequence: $\{2n^2 - 5n + 3\}$

2. Write the first six terms of the sequence: $\left\{n! + \dfrac{(n+1)!}{2^n}\right\}$

Note: The factorial operation, $n!$, is defined as follows: $n! = n(n-1)(n-2) \cdots 3 \cdot 2 \cdot 1$. In other words, you multiply the number by every positive integer smaller than that number. (For practice working with factorials, see Chapter 16.)

3. Write the first six terms of the sequence:
$\{(-1)^{n+1}3^n\}$

4. Write the first six terms of the sequence:
$\{(-1)^{n+1}n^2 + (-1)^n n^3\}$

5. Write the next two terms of this sequence, and then write a rule for determining all the terms of the sequence:
$\{1, -4, 9, -16, 25, -36, \ldots\}$

6. Write the next two terms of this sequence, and then write a rule for determining all the terms of the sequence:
$\left\{\dfrac{1}{2}, \dfrac{1}{4}, \dfrac{1}{8}, \dfrac{1}{16}, \dfrac{1}{32}, \dfrac{1}{64}, \ldots\right\}$

Differences and Multipliers: Working with Special Sequences

Two very special types of sequences are the arithmetic sequences and geometric sequences:

✔ **Arithmetic sequences:** These sequences always have terms that are a constant difference apart. For instance, the sequence $\{3, 8, 13, 18, 23, \ldots\}$ is an arithmetic sequence with terms that you can find by adding the difference, 5, to the preceding term.

Arithmetic sequences have a general term of $a_n = a_{n-1} + d$, where you can find the next term by adding the difference, d, to the preceding term, a_{n-1}. To find a particular term in the sequence — the nth term — you can also use the formula $a_n = a_1 + (n-1)d$, where a_1 is the first term in the sequence and d is the difference between the terms.

✔ **Geometric sequences:** These sequences have terms that are a constant multiplier or ratio apart. For instance, the sequence $\{1, 3, 9, 27, 81, \ldots\}$ is a geometric sequence with terms that you can find by multiplying the ratio, 3, by the preceding term.

Geometric sequences have a general term of $g_n = rg_{n-1}$, where you can find the next term by multiplying the ratio, r, by the preceding term, g_{n-1}. To find a particular term in the sequence — the nth term — you can also use the formula $g_n = g_1 r^{n-1}$, where g_1 is the first term in the sequence and r is the ratio.

Q. Find the 100th and 101st terms in the sequence beginning with −45, −43, −41, −39, . . . and also find a formula for all the terms in the sequence.

A. **Terms: 153 and 155; formula: {2*n* − 47}.** This sequence is arithmetic, because you can see a common difference of 2 between each pair of terms. The first term in the sequence is −45. To find a formula, use $a_n = a_1 + (n − 1)d$, replacing the a_1 with −45 and *d* with 2. Solving for the general term, $a_n = −45 + (n − 1) \cdot 2 = −45 + 2n − 2 = −47 + 2n$. Using the formula and substituting in the 100 for the *n* in a_n, you get $a_{100} = 2(100) − 47 = 200 − 47 = 153$. The 101st term is the next term, so just add 2 to 153 to get 155.

---

7. Find the 10th, 11th, and 12th terms of the arithmetic sequence whose first term is −4 and whose difference between the terms is 3.

Hint: Find the formula first.

8. Find the 10th, 11th, and 12th terms of the geometric sequence whose first term is 512 with a ratio of $\frac{3}{2}$.

Hint: Find the formula first.

---

9. Find the next three terms of the sequence beginning with 9, 15, 21, 27, . . . and write a formula for all the terms.

10. Find the next three terms of the sequence beginning with 1, −5, 25, −125, . . . and write a formula for all the terms.

Backtracking: Constructing Recursively Defined Sequences

A *recursively* defined sequence is one in which the successive terms are built from two or more of the preceding terms. One of the best-known recursively defined sequences is the Fibonacci sequence: 1, 1, 2, 3, 5, 8, 13, 21, 34, . . . — each term after the first two is the sum of the preceding two terms.

To define the Fibonacci sequence with the proper notation, you say that $a_1 = 1$, $a_2 = 1$, and $a_n = a_{n-2} + a_{n-1}$. The subscript notation says that a_{n-2} is two terms before the term being created and that a_{n-1} is the term right before the one being created.

Q. Determine the first eight terms of the sequence where $a_1 = 3$, $a_2 = 5$, and $a_n = 2u_{n-2} - a_{n-1}$.

A. **3, 5, 1, 9, –7, 25, –39, 89.** You already have the 3 and 5. The next term is $a_3 = 2a_{3-2} - a_{3-1} = 2a_1 - a_2 = 2(3) - 5 = 6 - 5 = 1$. The fourth term is $a_4 = 2a_{4-2} - a_{4-1} = 2a_2 - a_3 = 2(5) - 1 = 10 - 1 = 9$. The fifth term is

$a_5 = 2a_{5-2} - a_{5-1} = 2a_3 - a_4 = 2(1) - 9 = 2 - 9 = -7$. The sixth term is $a_6 = 2a_{6-2} - a_{6-1} = 2a_4 - a_5 = 2(9) - (-7) = 18 + 7 = 25$. The seventh term is $a_7 = 2a_{7-2} - a_{7-1} = 2a_5 - a_6 = 2(-7) - 25 = -14 - 25 = -39$. And the eighth term is $a_8 = 2a_{8-2} - a_{8-1} = 2a_6 - a_7 = 2(25) - (-39) = 50 + 39 = 89$.

---

11. Determine the first six terms of the sequence where $a_1 = 2$, $a_2 = 3$, and $a_n = 2a_{n-2} - 3a_{n-1}$.

12. Determine the first six terms of the sequence where $a_1 = 4$, $a_2 = 1$, and $a_n = a_{n-2} - (a_{n-1})^2$.

13. Determine the first six terms of the sequence where $a_1 = 2$, $a_2 = 4$, $a_3 = 7$, and $a_n = 3a_{n-3} + a_{n-1}$.

14. You're offered a job with a salary of $10,000 the first year and $20,000 the second year. After the second year, your salary will be the sum of the previous year's salary and 10 percent of the salary two years before that year. The formula for the salaries forms a sequence that's recursively defined. Find the first six years' salary amounts, using $a_1 = 10,000$, $a_2 = 20,000$, and $a_n = 0.1a_{n-2} + a_{n-1}$.

Using Summation Notation

A sequence is a list of things — usually numbers — and a series is the sum of a list of numbers. You can say, "Find the sum of the first six numbers in the sequence that has the general rule $a_n = n^2 - 3n + 4$." Or you can use *summation notation,* which tells you the same thing in a more cryptic (but quicker) fashion.

The summation notation for finding the sum of the first six terms of the sequence $a_n = n^2 - 3n + 4$ is $\sum_{n=1}^{6} (n^2 - 3n + 4)$. The big Σ stands for *summation* — it's the capital letter sigma in the Greek alphabet. The values below and above the sigma tell you where to start ($n = 1$) and stop (6), and the rule after the sigma, in the parentheses, tells you how to form the numbers to be added up. In this case, you add up $2 + 2 + 4 + 8 + 14 + 22$ to get 52.

Q. Find the sum of the series $\sum_{n=1}^{5} (n^2 + 3^n - 3)$.

A. **403.** Using the rule to list the terms to be added up, you get $(1 + 3 - 3) + (4 + 9 - 3) + (9 + 27 - 3) + (16 + 81 - 3) + (25 + 243 - 3) = 1 + 10 + 33 + 94 + 265 = 403$.

Q. Find the sum of the series $\sum_{n=1}^{50} \left(\frac{1}{2n-1} - \frac{1}{2n+1} \right)$.

A. $\frac{100}{101}$. At first, this problem may seem to be a bit daunting, with 50 terms to deal with and all those fractions. But after you list a few terms, you may see a pattern that can help you:

$$\left(\frac{1}{1} - \frac{1}{3}\right) + \left(\frac{1}{3} - \frac{1}{5}\right) + \left(\frac{1}{5} - \frac{1}{7}\right) + \left(\frac{1}{7} - \frac{1}{9}\right) + \cdots$$

$$= 1 + \left(-\frac{1}{3} + \frac{1}{3}\right) + \left(-\frac{1}{5} + \frac{1}{5}\right) + \left(-\frac{1}{7} + \frac{1}{7}\right) + \left(-\frac{1}{9} + \cdots\right)$$

The second term in each grouping is the opposite of the first term in the next grouping. All the opposites sum to 0. So, only the first term and the last term will be left. You can jump right to the final sum:

$$= 1 + \left(-\frac{1}{3} + \frac{1}{3}\right) + \left(-\frac{1}{5} + \frac{1}{5}\right) + \cdots + \left(-\frac{1}{97} + \frac{1}{97}\right) + \left(-\frac{1}{99} + \frac{1}{99}\right) - \frac{1}{101}$$

$$= 1 - \frac{1}{101} = \frac{101}{101} - \frac{1}{101} = \frac{100}{101}$$

---

15. Find the sum of the series $\displaystyle\sum_{n=1}^{4} \left(3^{n+1} - 4\right)$.

16. Find the sum of the series $\displaystyle\sum_{n=3}^{6} \left(n^2 - 2n - 1\right)$.

---

17. Find the sum of the series $\displaystyle\sum_{n=0}^{5} \left(\frac{n}{(n+1)^2}\right)$.

18. Find the sum of the series $\displaystyle\sum_{n=0}^{4} \left(\frac{n!}{n+1}\right)$.

Note: $n!$ means $n(n-1)(n-2)\cdots 3 \cdot 2 \cdot 1$. Note that by definition, $0! = 1$. For more info on factorial operations, see Chapter 16.

Finding Sums with Special Series

The sums of the terms in sequences have many practical applications, from performing calculations in the world of finance and banking to counting the seats in an amphitheater (or, naturally, impressing your date by determining the value of e to 100 decimal places). If the number of terms in the sequence is very large, the summing can get pretty awful, and the sums can get enormous.

Many commonly used sequences have series that have formulas for their sums. Algebraic and geometric series, for instance, have rules for their sums. Also, series consisting of terms that are squares of numbers or cubes of numbers have formulas for their sums. Refer to Chapter 18 if you want to see some of the special formulas that I use in this section.

Some formulas for the sums of series are in Figure 15-1.

Sequence Type	Rule for Sequence Terms	Formula for Sum of Terms
Arithmetic	$a_n = a_1 + (n-1)d$	$S_n = \dfrac{n}{2}[2a_1 + (n-1)d] = \dfrac{n}{2}[a_1 + a_n]$
Geometric	$g_n = g_1(r)^{n-1}$	$S_n = \dfrac{g_1(1-r^n)}{1-r}$
Converging geometric	$g_n = g_1(r)^{n-1}, |r| < 1$	$S_\infty = \dfrac{g_1}{1-r}$
Odd numbers	$a_n = 2n - 1$	$S_n = n^2$

Figure 15-1:
Sequences and their sum formulas.

EXAMPLE

Q. Find the sum of the series $2 + 5 + 8 + 11 + 14 + \ldots + 89$.

A. **1,365.** The terms in this series come from an arithmetic sequence of terms. The common difference is 3. You can use either formula from Figure 15-1 to find the sum, but in each, you need to know how many terms are in the list. Counting them one by one is a bit tedious and is likely to produce an error. Instead, create the rule for the general term. The terms are three apart, so you need $3n$ in your rule. Subtract 1 from a multiple of 3, and you get the terms. The rule is $3n - 1$. Set that rule equal to 89 to solve for n. If $3n - 1 = 89$, then $3n = 90$ and $n = 30$. This series contains 30 terms. Now, using the number of terms and the first and last terms in the formula, $S_n = \dfrac{30}{2}(2 + 89) = 15(91) = 1,365$.

Q. Find the sum of the series $1 + \dfrac{1}{4} + \dfrac{1}{16} + \dfrac{1}{64} + \dfrac{1}{256} + \ldots$; this problem asks for the sum of an infinite number of terms.

A. $\dfrac{4}{3}$. The terms in this series come from a geometric sequence of terms in which the ratio is a proper fraction — its absolute value is less than 1. You can use the general sum formula for a geometric series if you're adding up a certain number of terms — 10 or 20 or 50 of them. But you can also find the sum of all of them when the ratio of the geometric series is a proper fraction. The terms in a converging geometric series get so very small that their sum never exceeds a particular number — the series has a limit (see Chapter 8 for details on limits). Using the formula for an infinite sum on this series, you put the first term, 1, in the numerator and the ratio, $\dfrac{1}{4}$, in the denominator:

$$S_\infty = \dfrac{1}{1 - \dfrac{1}{4}} = \dfrac{1}{\dfrac{3}{4}} = \dfrac{4}{3}.$$

Q. Find the sum of the series $\frac{1}{4^5} + \frac{1}{4^6} + \frac{1}{4^7} + \frac{1}{4^8} + \ldots$

A. $\frac{1}{768}$. One way to approach this problem is to find the sum of the infinite series starting with

a first term, $\frac{1}{4^5} = \frac{1}{1,024}$, and using the formula where the ratio is $\frac{1}{4}$. You get $\dfrac{\frac{1}{1,024}}{1-\frac{1}{4}} = \dfrac{\frac{1}{1,024}}{\frac{3}{4}} =$

$\dfrac{1}{\cancel{1,024}_{256}} \cdot \dfrac{\cancel{4}}{3} = \dfrac{1}{768}$.

Another approach is to subtract the first four terms of a series with a ratio of $\frac{1}{4}$ from the infinite series. As you see, the answer is the same:

$$\sum_{n=0}^{\infty}\left(\frac{1}{4^n}\right) - \sum_{n=0}^{4}\left(\frac{1}{4^n}\right) = \frac{4}{3} - \frac{1\left(1-\left(\frac{1}{4}\right)^4\right)}{1-\frac{1}{4}} = \frac{4}{3} - \frac{\left(1-\frac{1}{1,024}\right)}{\frac{3}{4}}$$

$$= \frac{4}{3} - \frac{\frac{1,023}{1,024}}{\frac{3}{4}} = \frac{4}{3} - \frac{1,023}{1,024}\cdot\frac{4}{3}$$

$$= \frac{4}{3}\left(1-\frac{1,023}{1,024}\right) = \frac{4}{3}\left(\frac{1,024}{1,024} - \frac{1,023}{1,024}\right)$$

$$= \frac{\cancel{4}}{3}\left(\frac{1}{\cancel{1,024}_{256}}\right) = \frac{1}{768}$$

This technique doesn't look like a particularly good way to do the problem, but keep in mind that sometimes it's easier to subtract a few terms from the entire sum.

---

19. Find the sum of the series $1 + 2 + 3 + 4 + \cdots + 500$.

20. Find the sum of the series $1 + 2 + 4 + 8 + 16 + \cdots + 512$.

21. Find the sum of the series $-3 + 9 - 27 + 81 - 243 + \cdots + (-3)^{10}$.

22. Find the sum of the series $1 + 3 + 5 + 7 + 9 + \cdots + 99$.

23. Find the sum of the series $1 + \frac{2}{3} + \frac{4}{9} + \frac{8}{27} + \cdots$

24. Find the sum of the series $101 + 103 + 105 + \cdots + 1,001$.

Answers to Problems on Sequences and Series

The following are the answers to the practice problems presented earlier in this chapter.

1 Write the first six terms of the sequence: $\{2n^2 - 5n + 3\}$. The answer is **0, 1, 6, 15, 28, 45.**

Replace the n in the rule with 1 to get $2(1)^2 - 5(1) + 3 = 2 - 5 + 3 = 0$. Repeat the process, letting $n = 2, 3, 4, 5$, and 6 to determine the rest of the terms.

2 Write the first six terms of the sequence: $\left\{ n! + \dfrac{(n+1)!}{2^n} \right\}$. The answer is **2, 3.5, 9, 31.5, 142.5, 798.75.**

Replace the n in the rule with 1 to get $1! + \dfrac{(1+1)!}{2^1} = 1 + \dfrac{2!}{2} = 1 + \dfrac{2}{2} = 2$. Repeat the process, letting $n = 2, 3, 4, 5$, and 6 to determine the rest of the terms.

3 Write the first six terms of the sequence: $\{(-1)^{n+1}3^n\}$. The answer is **3, –9, 27, –81, 243, –729.**

Replace the n in the rule with 1 to get $(-1)^{1+1}3^1 = 1 \cdot 3 = 3$. Repeat the process, letting $n = 2, 3, 4, 5$, and 6 to determine the rest of the terms. Notice that even powers of –1 result in a positive term and that odd powers give you a negative term.

4 Write the first six terms of the sequence: $\{ (-1)^{n+1}n^2 + (-1)^n n^3 \}$. The answer is **0, 4, –18, 48, –100, 180.**

Replace the n in the rule with 1 to get $(-1)^{1+1}1^2 + (-1)^1 1^3 = 1 \cdot 1 + (-1) \cdot 1 = 1 - 1 = 0$. Repeat the process, letting $n = 2, 3, 4, 5$, and 6 to determine the rest of the terms.

5 Write the next two terms of this sequence, and then write a rule for determining all the terms of the sequence: $\{1, -4, 9, -16, 25, -36, \ldots\}$. The answer is **49, –64; $\{ (-1)^{n+1}n^2 \}$.**

The sequence consists of the squares of the numbers 1, 2, 3, 4, and so on with the signs alternating on the squares. When the square is an odd number, the sign is positive, so the exponent has to be even. You accomplish this adjustment by adding 1 to the number of the term in the exponent.

6 Write the next two terms of this sequence, and then write a rule for determining all the terms of the sequence: $\left\{ \dfrac{1}{2}, \dfrac{1}{4}, \dfrac{1}{8}, \dfrac{1}{16}, \dfrac{1}{32}, \dfrac{1}{64}, \ldots \right\}$. The answer is $\dfrac{1}{128}, \dfrac{1}{256}; \left\{ \dfrac{1}{2^n} \right\}$.

The sequence consists of powers of $\dfrac{1}{2}$ — in other words, powers of 2 in the denominator of a fraction. The powers shown are $2^1 = 2$, $2^2 = 4$, $2^3 = 8$, $2^4 = 16$, $2^5 = 32$, and $2^6 = 64$.

7 Find the 10th, 11th, and 12th terms of the arithmetic sequence whose first term is –4 and whose difference between the terms is 3. The answer is **23, 26, 29.**

Using $a_n = a_1 + (n - 1)d$, replace a_1 with –4 and d with 3 to get $a_n = -4 + (n - 1)3 = -4 + 3n - 3 = 3n - 7$. The rule for the sequence is $\{3n - 7\}$. Replace the n with 10 to get 23 for the 10th term. Then just add 3 for the 11th and 3 more for the 12th term.

8 Find the 10th, 11th, and 12th terms of the geometric sequence whose first term is 512 with a ratio of $\dfrac{3}{2}$. The answer is **19,683 then $\dfrac{59,049}{2}, \dfrac{177,147}{4}$.**

Using $g_n = g_1(r^{n-1})$, replace g_1 with 512 and r with to get $g_n = 512\left(\dfrac{3}{2}\right)^{n-1}$. Replacing the n with 10, you get $g_{10} = 512\left(\dfrac{3}{2}\right)^{10-1} = 2^9\left(\dfrac{3}{2}\right)^9 = \dfrac{2^9}{1} \cdot \dfrac{3^9}{2^9} = \dfrac{3^9}{1} = 19{,}683$. Multiply by the ratio to get the 11th term and then by the ratio, again, to get the 12th term.

9 Find the next three terms of the sequence beginning with 9, 15, 21, 27, . . . and write a formula for all the terms. The answer is **33, 39, 45; {6n + 3}**.

The sequence is arithmetic, because you can see a common difference between the terms: 6. The first term is 9, so using $a_n = a_1 + (n-1)d$, you determine that $a_n = 9 + (n-1)6 = 9 + 6n - 6 = 6n + 3$. Use this formula to find the next three terms — or just add 6 onto the 27, then 6 more and 6 more.

10 Find the next three terms of the sequence beginning with 1, –5, 25, –125, . . . and write a formula for all the terms. The answer is **625, –3,125, 15,625; $g_n = (-1)^{n+1}(5)^{n-1}$**.

This sequence is geometric, because you can see a constant multiple of –5 between the consecutive terms. Another way of writing the rule is $g_n = (-1)^{n-1}(5)^{n-1} = (-5)^{n-1}$. Use this rule to find the next three terms, or just multiply the –125 by –5 and then by –5 again and again.

11 Determine the first six terms of the sequence where $a_1 = 2$, $a_2 = 3$, and $a_n = 2a_{n-2} - 3a_{n-1}$. The answer is **2, 3, –5, 21, –73, 261**.

The rule has you perform operations on two consecutive terms to construct the next. Multiply the first 2 by 2 and subtract 3 times the 3 to get $2(2) - 3(3) = 4 - 9 = -5$. For the fourth term, multiply the 3 by 2 and subtract 3 times the –5 to get $2(3) - 3(-5) = 6 + 15 = 21$. Continue with the pattern to get the rest of the terms.

12 Determine the first six terms of the sequence where $a_1 = 4$, $a_2 = 1$, and $a_n = a_{n-2} - (a_{n-1})^2$. The answer is **4, 1, 3, –8, –61, –3,729**.

The rule has you perform operations on two consecutive terms to construct the next. Take the first term and subtract the square of the second from it. The third term is $4 - (1)^2 = 3$, the fourth term is $1 - (3)^2 = 1 - 9 = -8$, and so on.

13 Determine the first six terms of the sequence where $a_1 = 2$, $a_2 = 4$, $a_3 = 7$, and $a_n = 3a_{n-3} + a_{n-1}$. The answer is **2, 4, 7, 13, 25, 46**.

The rule has you perform operations on the first and third of three consecutive terms. You multiply the first term by 3 and add the third term to that product. The fourth term is $3(2) + 7 = 6 + 7 = 13$. The fifth term is $3(4) + 13 = 12 + 13 = 25$. The sixth term is $3(7) + 25 = 21 + 25 = 46$.

14 You're offered a job with a salary of $10,000 the first year and $20,000 the second year. After the second year, your salary will be the sum of the previous year's salary and 10 percent of the salary two years before that year. The formula for the salaries forms a sequence that's recursively defined. Find the first six years' salary amounts, using $a_1 = 10{,}000$, $a_2 = 20{,}000$, and $a_n = 0.1a_{n-2} + a_{n-1}$. The answer is **$10,000; $20,000; $21,000; $23,000; $25,100; $27,400**.

The rule has you perform operations on two consecutive salaries to determine the next. Multiply the first salary by 0.1 and add that amount to the second salary. The third year, the salary is $0.1(10{,}000) + 20{,}000 = 1{,}000 + 20{,}000 = 21{,}000$. The fourth year, the salary is $0.1(20{,}000) + 21{,}000 = 2{,}000 + 2{,}1000 = 23{,}000$. And so on.

15 Find the sum of the series $\sum_{n=1}^{4} (3^{n+1} - 4)$. The answer is **344**.

The sum of the four terms is $(5) + (27 - 4) + (81 - 4) + (243 - 4) = 5 + 23 + 77 + 239 = 344$.

16 Find the sum of the series $\sum_{n=3}^{6} (n^2 - 2n - 1)$. The answer is **46**.

Notice that the terms start with $n = 3$. The sum of the four terms is $(9 - 6 - 1) + (16 - 8 - 1) + (25 - 10 - 1) + (36 - 12 - 1) = 2 + 7 + 14 + 23 = 46$.

17 Find the sum of the series $\sum_{n=0}^{5} \left(\dfrac{n}{(n+1)^2} \right)$. The answer is $\dfrac{3,451}{3,600}$.

Notice that the terms start with $n = 0$. The sum of the six terms is

$$0 + \frac{1}{4} + \frac{2}{9} + \frac{3}{16} + \frac{4}{25} + \frac{5}{36} = \frac{900 + 800 + 675 + 576 + 500}{3,600} = \frac{3,451}{3,600}.$$

18 Find the sum of the series $\sum_{n=0}^{4} \left(\dfrac{n!}{n+1} \right)$. The answer is $\dfrac{127}{15}$.

You see that the terms start with $n = 0$. Also, remember that $0! = 1$, by definition. The sum of the five terms is $1 + \dfrac{1}{2} + \dfrac{2}{3} + \dfrac{3}{2} + \dfrac{24}{5} = \dfrac{30 + 15 + 20 + 45 + 144}{30} = \dfrac{254}{30} = \dfrac{127}{15}$.

19 Find the sum of the series $1 + 2 + 3 + 4 + \cdots + 500$. The answer is **125,250.**

The series is arithmetic with a first term of 1, a last term of 500, and a total of 500 terms. Using the formula, $S_n = \dfrac{n}{2} \left[a_1 + a_n \right] = \dfrac{500}{2} [1 + 500] = 250[501] = 125,250$.

20 Find the sum of the series $1 + 2 + 4 + 8 + 16 + \cdots + 512$. The answer is **1,023.**

The series is geometric with a first term of 1, a ratio of 2, and a tenth term of 512. Using the formula, $S_n = \dfrac{g_1(1 - r^n)}{1 - r} = \dfrac{1(1 - 2^{10})}{1 - 2} = \dfrac{1 - 1,024}{-1} = \dfrac{-1,023}{-1} = 1,023$. You determine that 512 is the tenth term by solving $1(2^{x-1}) = 512$ for x.

21 Find the sum of the series $-3 + 9 - 27 + 81 - 243 + \ldots + (-3)^{10}$. The answer is **44,286.**

The series is geometric with a first term of -3, a ratio of -3, and a total of 10 terms. Using the formula, $S_n = \dfrac{g_1(1 - r^n)}{1 - r} = \dfrac{-3\left(1 - (-3)^{10}\right)}{1 - (-3)} = \dfrac{-3(-59,048)}{4} = \dfrac{177,144}{4} = 44,286$.

22 Find the sum of the series $1 + 3 + 5 + 7 + 9 + \cdots + 99$. The answer is **2,500.**

The series is arithmetic with a first term of 1, a difference of 2, and 50 terms in the list. To determine the number of terms, you solve $2n - 1 = 99$ and find that $n = 50$. You can use the formula for the sum of the arithmetic series, or you can use the rule for the sum of n odd integers, which is n^2.

23 Find the sum of the series $1 + \dfrac{2}{3} + \dfrac{4}{9} + \dfrac{8}{27} + \ldots$ The answer is **3.**

The series is geometric with a ratio of $\dfrac{2}{3}$. Use the formula for an infinite series to get

$$S_\infty = \frac{1}{1 - \dfrac{2}{3}} = \frac{1}{\dfrac{1}{3}} = 3.$$

24 Find the sum of the series $101 + 103 + 105 + \cdots + 1,001$. The answer is **248,501.**

The series is arithmetic with a first term of 101, a difference of 2, and a total of 451 terms. You find the number of terms by solving $1,001 = 2n + 99$ for n. You can use the formula for the sum of the arithmetic series, or you can find n^2 and then subtract the terms from 1 through 99. Using the formula for the arithmetic series, $S_n = \dfrac{n}{2} \left[a_1 + a_n \right] = \dfrac{451}{2} [101 + 1,001] = \dfrac{451}{2} [1,102] = 248,501$.

Chapter 16

Everything You Ever Wanted to Know about Sets and Counting

A *set* is a mathematical entity that consists of a collection or group of objects. For instance, a set named A may consist of the first six prime numbers. You write $A = \{2, 3, 5, 7, 11, 13\}$. The set notation looks a bit like the notation you use to write the terms of a sequence (see Chapter 15), but with sets, the elements don't have to be in any particular order. You can just as easily write $A = \{3, 13, 11, 7, 5, 2\}$. The two sets here are the same.

In addition, sets don't require a mathematical rule for their elements. You can have $B = \{$horse, dog, 5, blue$\}$. The only thing the four elements in set B have in common is that they're in the same set. (Much like a family, a set can be a real mishmash.)

In this chapter, you become familiar with set notation and ways to perform operations such as union and intersection on sets. Then you count the number of elements in sets using the multiplication property, permutations, and combinations. I cover the binomial theorem, which is directly related to combinations, as well.

Writing the Elements of a Set from Rules or Patterns

You can describe the elements in a set in several different ways, but you usually want to choose the method that's quickest and most efficient and/or clearest to the reader. The two main methods for describing a set are roster and rule (or set-builder).

A *roster* is a list of the elements in a set. When the set doesn't include many elements, then this description works fine. If the set contains a lot of elements, you can use an ellipsis (. . .) if the pattern is obvious (a nasty word in mathematics). A rule works well when you find lots and lots of elements in the set.

Q. Use roster and rule notation to describe the set F, which consists of all the positive multiples of 5 that are less than 50.

A. **Roster notation: $F = \{5, 10, 15, 20, 25, 30, 35, 40, 45\}$; rule notation: $F = \{x \mid x = 5n$, where $n \in Z$ and $0 < n < 10\}$.** You read the rule as "Set F consists of all x's such that x is equal to five times n, where n is an integer and n is a number between zero and ten." (***Remember:*** The symbol Z stands for integers.) Of course, you don't have to write the elements in the set in order. You can just as easily write $F = \{10, 20, 30, 40, 45, 35, 25, 15, 5\}$.

Q. Write the elements of set C in roster form if $C = \{x \mid x = a^2$ and $x = b^3$, where $0 < a, b < 30\}$.

A. **$C = \{1, 64, 729\}$.** The rule for C is that x has to be a perfect square and a perfect cube. The bases of x (a and b) are positive numbers less than 30. The best way to approach this problem is to find all the squares of the numbers from 1 to 30 and then determine which are cubes: $C = \{1, 64, 729\}$. You get these elements because $1 = 1^2 = 1^3$, $64 = 8^2 = 4^3$, and $729 = 27^2 = 9^3$.

1. Write set A using roster notation if $A = \{x \mid x$ is odd, $x = 7n$, $0 < x < 70\}$.

2. Write set B using roster notation if $B = \{x \mid x = n^2 - 1$, $0 < n \leq 10\}$.

3. Write set C using a rule if $C = \{11, 21, 31, 41, 51, 61\}$.

4. Write set D using a rule if $D = \{1, 5, 9, 13, 17, 21, \ldots\}$.

Get Together: Combining Sets with Unions, Intersections, and Complements

Sets of elements can be combined or changed by using set operations. Much like addition or subtraction of real numbers, set operations are strictly defined to do something to the sets involved. The set operations are union, intersection, and complement:

✔ The *union* of two sets A and B is denoted $A \cup B$ and asks for all the elements in sets A and B — all of them together (without repeating any elements that they share).

✔ The *intersection* of the two sets A and B is denoted $A \cap B$ and asks for all the elements that A and B have in common. If the two sets have nothing in common, then your answer is the *empty set* or *null set*, written: { } or ∅.

✔ The *complement* of a set A, denoted A', asks for all the elements that *aren't* in the set but are in the universal set. The *universal set* is everything under consideration at the time. For instance, if you're working on sets that contain the letters of the English alphabet, then the universal set is all 26 letters.

Q. Given the sets $A = \{2, 4, 6, 8\}$, $B = \{4, 8, 16, 24, 32\}$, $C = \{3, 6, 9, 12, 15, 18, 21\}$, and the universal set is $U = \{x \mid 0 < x \le 32, x \in Z\} = \{1, 2, 3, 4, \ldots, 31, 32\}$, then find $A \cup B$, $A \cap B$, $B \cap C$, and C'.

A. The union of A and B, $A \cup B$, consists of all the elements in A and B both, so $A \cup B = \{2, 4, 6, 8, 16, 24, 32\}$. Notice that the 4 and 8 aren't repeated. The intersection of A and B, $A \cap B$, consists of all the elements the two sets share, so $A \cap B = \{4, 8\}$. The intersection of B and C, $B \cap C$, is all the elements that the two sets share, but the two sets have nothing in common, so $B \cap C = \varnothing$, the empty set. The complement of set C consists of everything that's in the universal set that's *not* in set C, so $C' = \{1, 2, 4, 5, 7, 8, 10, 11, 13, 14, 16, 17, 19, 20, 22, 23, 24, 25, 26, 27, 28, 29, 30, 31, 32\}$.

Q. Given the sets $D = \{0, 1, 2, 3, 4\}$, $E = \{0, 2, 6, 10\}$, $F = \{1, 3, 6, 10\}$, and $U = \{0, 1, 2, 3, \ldots, 10\}$, find the sets $(D \cap F)'$ and $(D \cup E) \cap (E \cup F)$.

A. $(D \cap F)' = \{0, 2, 4, 5, 6, 7, 8, 9, 10\}$; $(D \cup E) \cap (E \cup F) = \{0, 1, 2, 3, 6, 10\}$. The parentheses in these set operation problems work the same way as parentheses in algebraic expressions — you perform what's inside the parentheses first. To find the complement of the intersection of sets D and F, $(D \cap F)'$, you first find the intersection $D \cap F = \{1, 3\}$; then, referring back to the universal set, U, you find the complement: $(D \cap F)' = \{0, 2, 4, 5, 6, 7, 8, 9, 10\}$. The complement is everything *except* the 1 and 3 in the intersection. To find the intersection of the two unions, first find the two unions: $D \cup E = \{0, 1, 2, 3, 4, 6, 10\}$, and $E \cup F = \{0, 1, 2, 3, 6, 10\}$. You write the intersection — what the two results have in common — as $(D \cup E) \cap (E \cup F) = \{0, 1, 2, 3, 6, 10\}$.

5. Given the sets $A = \{0, 2, 4, 6, 8, \ldots, 20\}$, $B = \{0, 5, 10, 15, 20\}$, $C = \{7, 11, 17\}$, and the universal set $U = \{0, 1, 2, 3, 4, \ldots, 20\}$, find $A \cap B$.

6. Given the sets $A = \{0, 2, 4, 6, 8, \ldots, 20\}$, $B = \{0, 5, 10, 15, 20\}$, $C = \{7, 11, 17\}$, and the universal set $U = \{0, 1, 2, 3, 4, \ldots, 20\}$, find $A \cup B$.

7. Given the sets $A = \{0, 2, 4, 6, 8, \ldots, 20\}$, $B = \{0, 5, 10, 15, 20\}$, $C = \{7, 11, 17\}$, and the universal set $U = \{0, 1, 2, 3, 4, \ldots, 20\}$, find $B \cap C$.

8. Given the sets $A = \{0, 2, 4, 6, 8, \ldots, 20\}$, $B = \{0, 5, 10, 15, 20\}$, $C = \{7, 11, 17\}$, and the universal set $U = \{0, 1, 2, 3, 4, \ldots, 20\}$, find A'.

9. Given the sets $A = \{0, 2, 4, 6, 8, \ldots, 20\}$, $B = \{0, 5, 10, 15, 20\}$, $C = \{7, 11, 17\}$, and the universal set $U = \{0, 1, 2, 3, 4, \ldots, 20\}$, find $(A \cup C)$.

10. Given the sets $A = \{0, 2, 4, 6, 8, \ldots, 20\}$, $B = \{0, 5, 10, 15, 20\}$, $C = \{7, 11, 17\}$, and the universal set $U = \{0, 1, 2, 3, 4, \ldots, 20\}$, find $(A \cup B) \cap (B \cup C)$.

Multiplication Countdowns: Simplifying Factorial Expressions

Sets of elements have special operations used to combine them or change them. (The preceding section covers all those operations, if you need to refer back.) Another operation that's used with sets (but that isn't exclusive to sets) is *factorial*, denoted by the exclamation point. You use the factorial operation in the formulas used to count the number of elements in the union, intersection, or complement of sets. Factorials appear in the formulas you use to count the elements in sets that are really large. The upcoming sections "Counting on Permutations When Order Matters" and "Mixing It Up with Combinations" show you how the factorial works in the formulas.

The factorial operation, $n!$, is defined as $n! = n(n-1)(n-2)(n-3) \cdots 4 \cdot 3 \cdot 2 \cdot 1$. In other words, you multiply the number n, being operated upon, by every positive integer smaller than n. Some values of $n!$ are: $1! = 1$, $2! = 2$, $3! = 6$, $4! = 24$, $5! = 120$, $6! = 720$, and so on. You see that they're getting pretty big pretty fast.

One other factorial value that you need is $0! = 1$. You may think that's a typo. Nope. By definition, 0 factorial is equal to 1. It's one of those quirky things that mathematicians declare and make everyone use so that answers to problems come out right. People wanted the formulas for counting to be consistent for all the numbers used.

Simplifying factorials isn't difficult, but it isn't as easy as you may think at first glance. To simplify $\frac{6!}{3!}$, you can't just reduce the 6 and the 3. You have to look at all the factors involved in each factorial operation. Write out the factorials, and you get $\frac{6!}{3!} = \frac{6 \cdot 5 \cdot 4 \cdot 3 \cdot 2 \cdot 1}{3 \cdot 2 \cdot 1}$. Now reduce the like factors and simplify: $\frac{6!}{3!} = \frac{6 \cdot 5 \cdot 4 \cdot \cancel{3} \cdot \cancel{2} \cdot \cancel{1}}{\cancel{3} \cdot \cancel{2} \cdot \cancel{1}} = 6 \cdot 5 \cdot 4 = 120$.

Q. Simplify the factorial expression: $\frac{18!}{3!15!}$

A. **816.** First, write out the expansions of the factorials. But wait! (Notice that despite the exclamation point, the factorial doesn't work on the word *wait*.) Instead of writing out all the factors of 18!, just write 18! as $18 \cdot 17 \cdot 16 \cdot 15!$. You choose to stop with the 15 because of the 15! in the denominator. The 15! terms will cancel out, so don't bother to write out all those identical terms in both numerator and denominator:

$\frac{18!}{3!15!} = \frac{18 \cdot 17 \cdot 16 \cdot 15!}{3 \cdot 2 \cdot 1 \cdot 15!}$. Now divide out any other common factors and simplify:

$\frac{18!}{3!15!} = \frac{\overset{3}{\cancel{18}} \cdot 17 \cdot 16 \cdot \cancel{15!}}{\cancel{3} \cdot 2 \cdot 1 \cdot \cancel{15!}} = 3 \cdot 17 \cdot 16 = 816$.

11. Simplify the expression: $\frac{8!}{4!}$

12. Simplify the expression: $\frac{52!}{50!}$

13. Simplify the expression: $\frac{5!}{2!3!}$

14. Simplify the expression: $\frac{20!}{5!15!}$

Checking Your Options: Using the Multiplication Property

When you want to count up how many things are in a set, you have quite a few options. When the set contains too many elements to count accurately, you look for some sort of pattern or rule to help out. In this section, you practice the multiplication property. In the next two sections, you count the elements using permutations and combinations. Each method has its place in the world of mathematical counting. You can count on it!

If you can do task one in m_1 ways, task two in m_2 ways, task three in m_3 ways, and so on, then you can perform all the tasks in a total of $m_1 \cdot m_2 \cdot m_3 \ldots$ ways.

Q. How many ways can you fly from San Francisco to New York City, stopping in Denver, Chicago, and Buffalo, if the website offers four ways to fly from San Francisco to Denver, six ways to fly from Denver to Chicago, two ways to fly from Chicago to Buffalo, and three ways to fly from Buffalo to New York City?

A. **144.** Multiply $4 \cdot 6 \cdot 2 \cdot 3 = 144$. This method doesn't tell you what all the routes are; it just tells you how many are possible so you know when you've listed all of them. (Better get to work on that.)

Q. How many ways can you write a password if the first symbol has to be a digit from 1 to 9; the second, third, and fourth symbols have to be letters of the English alphabet; and the last symbol has to be from the set {!, @, #, $, %, ^, &, *, +}?

A. **1,423,656.** You multiply $9 \cdot 26 \cdot 26 \cdot 26 \cdot 9 = 1,423,656$. This system allows a lot of passwords, but most institutions make you use eight or more characters, which makes the number of possibilities even greater.

15. If you have to take one class in each subject, how many different course loads can you create if you have a choice of four math classes, three history classes, eight English classes, and five science classes?

16. How many different ice-cream sundaes can you create if you have a choice of five ice-cream flavors, three sauces, and five sprinkled toppings if you choose one of each type?

17. How many different automobiles can you order if you have a choice of six colors, four interiors, two trim options, three warranties, and two types of seats?

18. How many different dinners can you order if you have a choice of 12 appetizers, 8 entrees, 5 potatoes, 6 desserts, and a choice of soup or salad?

Counting on Permutations When Order Matters

Permutations involve taking a specific number of items from an available group or set and seeing how many different ways the items can be selected and then arranged. For instance, if you choose three letters from the set {*a, r, s, t*} and arrange them as many ways as possible, you get the arrangements of {*a, r, s*}: *ars, asr, ras, rsa, sar,* and *sra;* the arrangements of {*a, r, t*}: *art, atr, rat, rta, tar,* and *tra;* and the arrangements of {*a, s, t*}: *ast, ats, sat, sta, tas,* and *tsa;* and the arrangements of {*r, s, t*}: *rst, rts, srt, str, trs,* and *tsr.* The number of permutations is 24.

Note: If you lose track of how to figure out all the "words," you can list all these arrangements using a *tree*. You can find a discussion of trees in *Algebra II For Dummies* (Wiley) or in other algebra books. Without a tree, just figure out a way to do a systematic listing.

You can find the number of permutations of *n* things taken *r* at a time with the formula $P(n, r) = \dfrac{n!}{(n-r)!}$. The *n* is the grouping or set that you're pulling items from. The *r* is how many of those items you're taking at a time. The notation $P(n, r)$ or $_nP_r$ is standard notation for indicating permutations.

Q. How many permutations are possible if you choose three letters from a set of four? (This problem is the one I describe at the beginning of this section.)

A. **24.** Using the formula, $P(4, 3) = \dfrac{4!}{(4-3)!} = \dfrac{4!}{1!} = \dfrac{4 \cdot 3 \cdot 2 \cdot 1}{1} = 24$.

Q. How many arrangements are possible if you choose any three letters from the English alphabet and then take any three digits from the digits 0 through 9 and use them for a password (assume that none is repeated)?

A. **11,232,000.** First, use the formula to find the number of arrangements of letters: $P(26, 3) = \dfrac{26!}{(26-3)!} = \dfrac{26!}{23!} = \dfrac{26 \cdot 25 \cdot 24 \cdot 23!}{23!} = 26 \cdot 25 \cdot 24 = 15{,}600$. Then find the number of arrangements of digits: $P(10, 3) = \dfrac{10!}{(10-3)!} = \dfrac{10!}{7!} = \dfrac{10 \cdot 9 \cdot 8 \cdot 7!}{7!} = 10 \cdot 9 \cdot 8 = 720$. Finally, using the multiplication from the preceding section, "Checking Your Options: Using the Multiplication Property," multiply the two answers together: $15{,}600 \cdot 720 = 11{,}232{,}000$.

---

19. How many arrangements ("words") are possible using three of the letters from the word *stare?*

20. How many arrangements ("words") are possible using all five of the letters from the word *stare?*

21. How many different license plates can you form using three letters from the English alphabet (except the letters *O* and *I* and not repeating any), two digits from the digits 1 through 9 (without repetition), and then two letters chosen from the first seven letters in the alphabet (not two of the same letter)?

22. You have six blue books, five red books, and ten green books, and you decide to put four of each color on a bookshelf, keeping the same colors together. How many arrangements are possible?

Mixing It Up with Combinations

Combinations are another way of counting items. This time, you select some items from a larger group, but you don't care what order they come in. For instance, if you select 6 lottery numbers from a listing of 54, which number was chosen first or second or third doesn't matter — just that you chose them.

You can find the number of combinations or ways to choose *r* elements from a set containing *n* elements with the formula $C(n, r) = \begin{pmatrix} n \\ r \end{pmatrix} = \frac{n!}{r!(n-r)!}$. The two different notations and ${}_nC_r$ are commonly used to indicate that you're finding the number of combinations. Notice that this formula looks very much like the formula for permutations. The extra factor in the denominator makes the total number smaller.

Q. Find the number of ways (combinations) of selecting 6 lottery numbers from 54 choices.

A. **25,827,165.** (At \$1 per entry, no wonder your chances are so slim.) Here's the math:

$$C(54, 6) = \begin{pmatrix} 54 \\ 6 \end{pmatrix} = \frac{54!}{6!(54-6)!} = \frac{54!}{6!48!} = \frac{54 \cdot 53 \cdot 52 \cdot 51 \cdot 50 \cdot 49 \cdot 48!}{6 \cdot 5 \cdot 4 \cdot 3 \cdot 2 \cdot 1 \cdot 48!}$$

$$= \frac{{}^9 54 \cdot 53 \cdot {}^{13}52 \cdot {}^{17}51 \cdot {}^5 50 \cdot 49 \cdot 48!}{6 \cdot 5 \cdot 4 \cdot 3 \cdot 2 \cdot 1 \cdot 48!} = 9 \cdot 53 \cdot 13 \cdot 17 \cdot 5 \cdot 49$$

$$= 25,827,165$$

Each of the factors in the denominator divides into a factor in the numerator, so the result is always a whole number.

Q. At a restaurant, the specialty platter includes a choice of any two of the five entrees and any three of the ten sides. How many different platters are possible?

A. **1,200.** Use $C(5, 2)$ to choose the entrees and $C(10, 3)$ to choose the sides. Then multiply the two results together using the multiplication property from earlier in this chapter:

$$C(5, 2) \cdot C(10, 3) = \frac{5!}{2!3!} \cdot \frac{10!}{3!7!} = \frac{5 \cdot 4 \cdot 3!}{2 \cdot 1 \cdot 3!} \cdot \frac{10 \cdot 9 \cdot 8 \cdot 7!}{3 \cdot 2 \cdot 1 \cdot 7!}$$

$$= \frac{5 \cdot {}^2 4}{2 \cdot 1} \cdot \frac{{}^5 10 \cdot {}^3 9 \cdot 8}{3 \cdot 2 \cdot 1} = 5 \cdot 2 \cdot 5 \cdot 3 \cdot 8 = 1,200$$

You find that 1,200 different platters are possible.

23. You can invite 9 of your 20 best friends to your birthday party. How many different combinations of friends are possible?

24. A lottery has you choose 8 numbers from 60. How many different winning combinations are possible?

25. You're selecting a committee of eight. Four are to be chosen from a group of 10 men, and 4 are to be chosen from a group of 20 women. How many different committees can you have?

26. You're going to sample 5 fruit cups out of the total shipment of 400. If those 5 cups are okay, then you can assume that the whole shipment is okay. How many different ways can you choose 5 fruit cups out of 400?

Raising Binomials to Powers: Investigating the Binomial Theorem

A *binomial* is a mathematical expression that has two terms. In algebra, people frequently raise binomials to powers to complete computations.

The *binomial theorem* says that if a and b are real numbers and n is a positive integer, then

$$(a+b)^n = a^n + na^{n-1}b^1 + \frac{n(n-1)}{2}a^{n-2}b^2 + L + na^1b^{n-1} + b^n$$

$$= \binom{n}{0}a^n + \binom{n}{1}a^{n-1}b^1 + L + \binom{n}{k}a^{n-k}b^k + L \binom{n}{n-1}a^1b^{n-1} + \binom{n}{n}b^n$$

You can see the rule here, in the second line, in terms of the coefficients that are created using combinations. The powers on a start with n and decrease until the power is 0 in the last term. That's why you don't see an a in the last term — it's really a^0. The powers on b increase from b^0 in the first term until the last term, where it's b^n. Notice that the power of b matches k in the combination.

Q. Use the binomial theorem to determine the coefficients of the expansion of $(x + y)^5$. Then write the expansion.

A. $x^5 + 5x^4y + 10x^3y^2 + 10x^2y^3 + 5xy^4 + y^5$. Using the format from the binomial theorem, the coefficients of the expansion are $\binom{5}{0}, \binom{5}{1}, \binom{5}{2}, \binom{5}{3}, \binom{5}{4}, \binom{5}{5} = 1, 5, 10, 10, 5, 1$.

Place those coefficients along a horizontal line, leaving room after each coefficient for the powers of x and y. You have something that looks like: 1 5 10 10 5 1. Now write decreasing powers of x after each coefficient, giving you $1x^5$ $5x^4$ $10x^3$ $10x^2$ $5x^1$ $1x^0$. The last power of x, the x^0, is just a 1, so you don't write it into the term. Now write increasing powers of y after each x term and on the last term, and add the terms together (the first power is 0, so you don't need a y after the first x term). You now get $1x^5 + 5x^4y^1 + 10x^3y^2 + 10x^2y^3 + 5x^1y^4 + 1y^5$. Notice that the sum of the exponents on the variables in each term is 5.

---

27. Expand $(x + y)^6$ using the binomial theorem.

28. Expand $(x - y)^4$ using the binomial theorem.

Answers to Problems on Sets and Counting

The following are the answers to the practice problems presented earlier in this chapter.

1 Write set A using roster notation if $A = \{x \mid x$ is odd, $x = 7n$, $0 < x < 70\}$. The answer is **{7, 21, 35, 49, 63}.**

According to the rule, you want numbers that are odd, multiples of 7, and between 0 and 70.

2 Write set B using roster notation if $B = \{x \mid x = n^2 - 1, 0 < n \leq 10\}$. The answer is **{0, 3, 8, 15, 24, 35, 48, 63, 80, 99}.**

The numbers in this set are all 1 less than a perfect square — and the perfect squares are all integers between 0 and 10.

3 Write set C using a rule if $C = \{11, 21, 31, 41, 51, 61\}$. The answer is **$C = \{x \mid x = 10n + 1, 1 \leq n \leq 6\}$.**

Another way of writing the rule is $C = \{x \mid x = 10n + 11, 0 \leq n \leq 5\}$. Notice that in this second rule, the values of n are different — they begin and end in different places — and the constant 11 is different. What's alike in these two rules is that the n is multiplied by 10, keeping the terms 10 units apart.

4 Write set D using a rule if $D = \{1, 5, 9, 13, 17, 21, \dots\}$. The answer is **$D = \{x \mid x = 4n + 1, n \geq 0\}$.**

The rule allows the set to be infinite — the number of terms has no end.

5 Given the sets $A = \{0, 2, 4, 6, 8, \dots, 20\}$, $B = \{0, 5, 10, 15, 20\}$, $C = \{7, 11, 17\}$, and the universal set $U = \{0, 1, 2, 3, 4, \dots, 20\}$, find $A \cap B$. The answer is **{0, 10, 20}.**

The sets A and B share only these three elements.

6 Given the sets $A = \{0, 2, 4, 6, 8, \dots, 20\}$, $B = \{0, 5, 10, 15, 20\}$, $C = \{7, 11, 17\}$, and the universal set $U = \{0, 1, 2, 3, 4, \dots, 20\}$, find $A \cup B$. The answer is **{0, 2, 4, 5, 6, 8, 10, 12, 14, 15, 16, 18, 20}.**

The union of A and B contains everything from A — all the even numbers from 0 to 20 — and everything from B — the multiples of 5 from 0 to 20. Essentially, you just list the even numbers and insert the 5 and 15 from set B. The 0 and 10 and 20 are already accounted for.

7 Given the sets $A = \{0, 2, 4, 6, 8, \dots, 20\}$, $B = \{0, 5, 10, 15, 20\}$, $C = \{7, 11, 17\}$, and the universal set $U = \{0, 1, 2, 3, 4, \dots, 20\}$, find $B \cap C$. The answer is $\varnothing$ **or { }.**

The solution is the empty set, because sets B and C have no terms in common.

8 Given the sets $A = \{0, 2, 4, 6, 8, \dots, 20\}$, $B = \{0, 5, 10, 15, 20\}$, $C = \{7, 11, 17\}$, and the universal set $U = \{0, 1, 2, 3, 4, \dots, 20\}$, find A'. The answer is **{1, 3, 5, 7, \dots, 19}.**

The complement of set A, the even numbers from 0 to 20, is a set that has all the odd numbers between 0 and 20.

9 Given the sets $A = \{0, 2, 4, 6, 8, \dots, 20\}$, $B = \{0, 5, 10, 15, 20\}$, $C = \{7, 11, 17\}$, and the universal set $U = \{0, 1, 2, 3, 4, \dots, 20\}$, find $(A \cup C)'$. The answer is **{1, 3, 5, 9, 13, 15, 19}.**

You first find the union of sets A and C: $A \cup C = \{0, 2, 4, 6, 7, 8, 10, 11, 12, 14, 16, 17, 18, 20\}$. The complement of that union is everything in the universal set, U, that isn't in the set $A \cup C$. So, $(A \cup C)' = \{1, 3, 5, 9, 13, 15, 19\}$.

10 Given the sets $A = \{0, 2, 4, 6, 8, \dots, 20\}$, $B = \{0, 5, 10, 15, 20\}$, $C = \{7, 11, 17\}$, and the universal set $U = \{0, 1, 2, 3, 4, \dots, 20\}$, find $(A \cup B) \cap (B \cup C)$. The answer is **{0, 5, 10, 15, 20}.**

First find the union $A \cup B$ and the other union $B \cup C$. $A \cup B = \{0, 2, 4, 5, 6, 8, 10, 12, 14, 15, 16, 18, 20\}$, and $B \cup C = \{0, 5, 7, 10, 11, 15, 17, 20\}$. The intersection of these two resulting sets is $(A \cup B) \cap (B \cup C) = \{0, 5, 10, 15, 20\} = B$. The result is the original set B.

11 Simplify the expression $\frac{8!}{4!}$. The answer is **1,680.**

Expand the numerator, and leave the denominator as 4!. Then reduce and simplify:

$\frac{8!}{4!} = \frac{8 \cdot 7 \cdot 6 \cdot 5 \cdot 4!}{4!} = 8 \cdot 7 \cdot 6 \cdot 5 = 1,680 \cdot$

12 Simplify the expression $\frac{52!}{50!}$. The answer is **2,652.**

Expand the numerator, and leave the denominator as 50!. Then reduce and simplify:

$\frac{52!}{50!} = \frac{52 \cdot 51 \cdot 50!}{50!} = 52 \cdot 51 = 2,652.$

13 Simplify the expression $\frac{5!}{2!3!}$. The answer is **10.**

Expand the numerator and the first factor in the denominator. Reduce the common factors and simplify: $\frac{5!}{2!3!} = \frac{5 \cdot 4 \cdot 3!}{2 \cdot 1 \cdot 3!} = \frac{5 \cdot \overset{2}{4}}{2 \cdot 1} = 5 \cdot 2 = 10 \cdot$

14 Simplify the expression $\frac{20!}{5!15!}$. The answer is **15,504.**

Expand the numerator and the first factor in the denominator. Reduce the common factors and simplify: $\frac{20!}{5!15!} = \frac{20 \cdot 19 \cdot 18 \cdot 17 \cdot 16 \cdot 15!}{5 \cdot 4 \cdot 3 \cdot 2 \cdot 1 \cdot 15!} = \frac{\overset{1}{20} \cdot 19 \cdot \overset{3}{18} \cdot 17 \cdot 16}{5 \cdot 4 \cdot 3 \cdot 2 \cdot 1} = 19 \cdot 3 \cdot 17 \cdot 16 = 15,504 \cdot$

15 If you have to take one class in each subject, how many different course loads can you create if you have a choice of four math classes, three history classes, eight English classes, and five science classes? The answer is **480.**

Multiply: $4 \cdot 3 \cdot 8 \cdot 5 = 480.$

16 How many different ice-cream sundaes can you create if you have a choice of five ice-cream flavors, three sauces, and five sprinkled toppings if you choose one of each type? The answer is **75.**

Multiply: $5 \cdot 3 \cdot 5 = 75.$

17 How many different automobiles can you order if you have a choice of six colors, four interiors, two trim options, three warranties, and two types of seats? The answer is **288.**

Multiply: $6 \cdot 4 \cdot 2 \cdot 3 \cdot 2 = 288.$

18 How many different dinners can you order if you have a choice of 12 appetizers, 8 entrees, 5 potatoes, 6 desserts, and a choice of soup or salad? The answer is **5,760.**

Multiply: $12 \cdot 8 \cdot 5 \cdot 6 \cdot 2 = 5,760.$ Don't forget that *soup or salad* is two choices for that selection.

19 How many arrangements ("words") are possible using three of the letters from the word *stare?* The answer is **60.**

Use the permutation formula $P(5, 3)$. Simplifying, $P(5, 3) = \frac{5!}{(5-3)!} = \frac{5 \cdot 4 \cdot 3 \cdot 2!}{2!} = 5 \cdot 4 \cdot 3 = 60.$

20 How many arrangements ("words") are possible using all five of the letters from the word *stare?* The answer is **120.**

Use the permutation formula $P(5, 5)$. Simplifying,

$P(5, 5) = \frac{5!}{(5-5)!} = \frac{5 \cdot 4 \cdot 3 \cdot 2 \cdot 1}{0!} = \frac{5 \cdot 4 \cdot 3 \cdot 2 \cdot 1}{1} = 5! = 120.$

21 How many different license plates can you form using three letters from the English alphabet (except the letters *O* and *I* and not repeating any), two digits from the digits 1 through 9 (without repetition), and then two letters chosen from the first seven letters in the alphabet (not two of the same letter)? The answer is **36,723,456.**

Use three different permutations all multiplied together. For the first three letters, use $P(24, 3)$. The two digits use $P(9, 2)$. And the last two letters use $P(7, 2)$:

$$P(24, 3) \cdot P(9, 2) \cdot P(7, 2) = \frac{24!}{(24-3)!} \cdot \frac{9!}{(9-2)!} \cdot \frac{7!}{(7-2)!}$$

$$= \frac{24!}{21!} \cdot \frac{9!}{7!} \cdot \frac{7!}{5!} = \frac{24!}{21!} \cdot \frac{9!}{5!}$$

$$= \frac{24 \cdot 23 \cdot 22 \cdot 21!}{21!} \cdot \frac{9 \cdot 8 \cdot 7 \cdot 6 \cdot 5!}{5!}$$

$$= 24 \cdot 23 \cdot 22 \cdot 9 \cdot 8 \cdot 7 \cdot 6 = 36,723,456$$

22 You have six blue books, five red books, and ten green books, and you decide to put four of each color on a bookshelf, keeping the same colors together. How many arrangements are possible? The answer is **1,306,368,000**.

Use four different permutations all multiplied together. For the blue books, use $P(6, 4)$; for the red books, use $P(5, 4)$; and for the green books, use $P(10, 4)$. You then have to account for what order the three colors are going to be in. Use $P(3, 3)$. The books will be ordered within their colors and the groups of colors will be ordered. Whew! Here's what the work looks like:

$$P(6, 4) \cdot P(5, 4) \cdot P(10, 4) \cdot P(3, 3) = \frac{6!}{(6-4)!} \cdot \frac{5!}{(5-4)!} \cdot \frac{10!}{(10-4)!} \cdot \frac{3!}{(3-3)!}$$

$$= \frac{6!}{2!} \cdot \frac{5!}{1!} \cdot \frac{10!}{6!} \cdot \frac{3!}{0!}$$

$$= \frac{5! \cdot 10! \cdot 3!}{2! \cdot 1 \cdot 1}$$

$$= \frac{5! \cdot 10! \cdot 3 \cdot 2!}{2!} = 5!10!3$$

$$= 120 \cdot 3,628,800 \cdot 3$$

$$= 1,306,368,000$$

23 You can invite 9 of your 20 best friends to your birthday party. How many different combinations of friends are possible? The answer is **167,960**.

This is a combination of 20 items in which you choose 9, $C(20, 9)$. Solving,

$$C(20, 9) = \frac{20!}{9!(20-9)!} = \frac{20!}{9!11!}$$

$$= \frac{20 \cdot 19 \cdot 18 \cdot 17.^2 16.^5 15.^2 14 \cdot 13.^2 12 \cdot 11!}{9 \cdot 8 \cdot 7 \cdot 6 \cdot 5 \cdot 4 \cdot 3 \cdot 2 \cdot 1 \cdot 11!}$$

$$= 19 \cdot 17 \cdot 2 \cdot 5 \cdot 2 \cdot 13 \cdot 2 = 167,960$$

You can reduce the fraction in lots of ways — you have many combinations of the factors. But in any case, you should end up with all 1s in the denominator and numbers to multiply in the numerator.

24 A lottery has you choose 8 numbers from 60. How many different winning combinations are possible? The answer is **2,558,620,845**.

It would be a wonder if anyone ever won this lottery! This problem involves a combination of 60 numbers, choosing 8. Solving,

$$C(60, 8) = \frac{60!}{8!(60-8)!} = \frac{60!}{8!52!}$$

$$= \frac{60 \cdot 59 \cdot^{29} 58 \cdot 57 \cdot 56 \cdot 55 \cdot^9 54 \cdot 53 \cdot 52!}{8 \cdot 7 \cdot 6 \cdot 5 \cdot 4 \cdot 3 \cdot 2 \cdot 1 \cdot 52!}$$

$$= 59 \cdot 29 \cdot 57 \cdot 55 \cdot 9 \cdot 53 = 2,558,620,845$$

25 You're selecting a committee of eight. Four are to be chosen from a group of 10 men, and 4 are to be chosen from a group of 20 women. How many different committees can you have? The answer is **1,017,450.**

Multiply $C(10, 4)$ times $C(20, 4)$. You're simply multiplying the number of ways to choose four men times the number of ways to choose four women:

$$C(10, 4) \cdot C(10, 4) = \frac{10!}{4!(10-4)!} \cdot \frac{20!}{4!(20-4)!}$$

$$= \frac{10 \cdot 9 \cdot 8 \cdot 7 \cdot \cancel{6!}}{4 \cdot 3 \cdot 2 \cdot 1 \cdot \cancel{6!}} \cdot \frac{20 \cdot 19 \cdot 18 \cdot 17 \cdot \cancel{16!}}{4 \cdot 3 \cdot 2 \cdot 1 \cdot \cancel{16!}}$$

$$= \frac{10 \cdot {}^3\cancel{9} \cdot \cancel{8} \cdot 7}{\cancel{4} \cdot \cancel{3} \cdot \cancel{2}} \cdot \frac{{}^5\cancel{20} \cdot 19 \cdot {}^3\cancel{18} \cdot 17}{\cancel{4} \cdot \cancel{3} \cdot \cancel{2}}$$

$$= 10 \cdot 3 \cdot 7 \cdot 5 \cdot 19 \cdot 3 \cdot 17 = 1,017,450$$

26 You're going to sample 5 fruit cups out of the total shipment of 400. If those 5 cups are okay, then you can assume that the whole shipment is okay. How many different ways can you choose 5 fruit cups out of 400? The answer is **83,218,600,080.**

Solve $C(400, 5)$:

$$\frac{400!}{5!(400-5)!} = \frac{400!}{5!395!} = \frac{{}^{10}\cancel{400} \cdot {}^{133}\cancel{399} \cdot 398 \cdot 397 \cdot 396 \cdot \cancel{395!}}{\cancel{5} \cdot \cancel{4} \cdot \cancel{3} \cdot \cancel{2} \cdot 1 \cdot \cancel{395!}}$$

$$= 10 \cdot 133 \cdot 398 \cdot 397 \cdot 396$$

$$= 83,218,600,080$$

Some scientific calculators have trouble with a number this large and may go into scientific notation for an answer. You may see something like $8.32186 \cdot 10^{10}$.

27 Expand $(x + y)^6$ using the binomial theorem. The answer is $x^6 + 6x^5y + 15x^4y^2 + 20x^3y^3 + 15x^2y^4 + 6xy^5 + y^6$.

Using the binomial theorem, the coefficients are
$\binom{6}{0}, \binom{6}{1}, \binom{6}{2}, \binom{6}{3}, \binom{6}{4}, \binom{6}{5}, \binom{6}{6} = 1, 6, 15, 20, 15, 6, 1$. First write the coefficients, leaving space for the decreasing powers of x: 1 6 15 20 15 6 1. Now write in the decreasing powers of x: $1x^6$ $6x^5$ $15x^4$ $20x^3$ $15x^2$ $6x$ 1. Put in increasing powers of y and add the terms together: $x^6 + 6x^5y + 15x^4y^2 + 20x^3y^3 + 15x^2y^4 + 6xy^5 + y^6$.

28 Expand $(x - y)^4$ using the binomial theorem. The answer is $x^4 - 4x^3y + 6x^2y^2 - 4xy^3 + 1$.

Using the binomial theorem, the coefficients are $\binom{4}{0}, \binom{4}{1}, \binom{4}{2}, \binom{4}{3}, \binom{4}{4} = 1, 4, 6, 4, 1$. Write out the coefficients, leaving space for the decreasing powers of x: 1 4 6 4 1. Now write in the decreasing powers of x: $1x^4$ $4x^3$ $6x^2$ $4x$ 1. The y term is negative, so the powers of y have to include the negative sign when you write in the increasing powers of y: $1x^4(-y)^0$ $4x^3(-y)^1$ $6x^2(-y)^2$ $4x(-y)^3$ $1(-y)^4$. The even powers of y turn out to be positive, and the odd powers are negative. This situation makes the terms in the expansion alternate from positive to negative and back again: $1x^4 - 4x^3y + 6x^2y^2 - 4xy^3 + 1y^4$.

Part V
The Part of Tens

In this part . . .

- ✔ Recognize what basic graphs have in common and what makes them different.

- ✔ Work with sums of arithmetic and geometric series, and use infinite sums to compute familiar mathematical values.

Chapter 17

Ten Basic Graphs

Graphing is one way of getting the characteristics of a function out there for everyone to see. The basic graphs are just that — basic. They're centered at the origin and aren't expanded or shrunken or jostled about. You can alter the basic graphs by performing translations to the left or right or up or down.

Chapter 4 presents the nuts and bolts of graphing, and you can refer to the earlier chapters in this book for more-specific information on how to change the looks of the basic graphs. Right now, though, sit back and take a look at some of these elegant images.

Putting Polynomials in Their Place

The graph of a polynomial function is a smooth curve that may or may not change direction, depending on its degree. The two simplest polynomials are the quadratic, $y = x^2$ (see Figure 17-1), and the cubic, $y = x^3$ (see Figure 17-2). They both go through the origin and the point (1, 1).

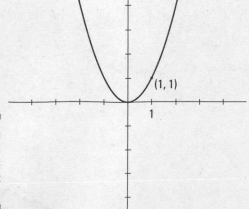

Figure 17-1:
The function $y = x^2$ has its vertex at the origin.

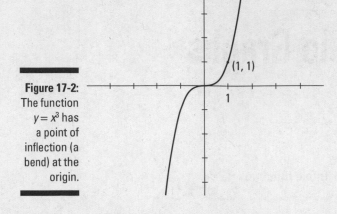

Figure 17-2:
The function
$y = x^3$ has
a point of
inflection (a
bend) at the
origin.

Lining Up Front and Center

Just two points determine a unique line. This statement means that only one line can go through any two designated points.

Lines can have *x*- and *y*-intercepts — where the lines cross the axes; the *slope* of a line tells whether it rises or falls and how steeply this happens. Chapter 4 has lots of information on lines, if you want to know more. As Figure 17-3 shows, the graph of the line $y = x$ goes diagonally through the first and third quadrants. The slope is 1, and the line goes through the point (1, 1). The only intercept of this line is the origin.

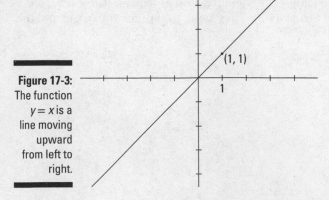

Figure 17-3:
The function
$y = x$ is a
line moving
upward
from left to
right.

Being Absolutely Sure with Absolute Value

The absolute value function $y = |x|$ has a characteristic V shape (see Figure 17-4). The V is typical of most absolute value equations with linear terms. The only intercept of this basic absolute value graph is the origin, and the function goes through the point (1, 1). Flip to Chapter 1 for more info on absolute value.

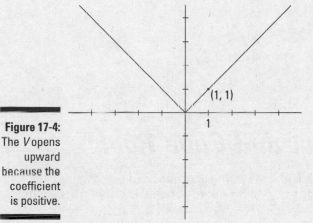

Figure 17-4:
The V opens upward because the coefficient is positive.

(1, 1)

1

Graphing Reciprocals of x and x²

The graphs of $y = \frac{1}{x}$ and $y = \frac{1}{x^2}$ both have vertical asymptotes of $x = 0$ and horizontal asymptotes of $y = 0$ (see Figures 17-5 and 17-6). The asymptotes are actually the x- and y-axes. (You can find more information on asymptotes in Chapter 8.) Each curve goes through the point (1, 1), and each curve exhibits symmetry. As you can see in the figures, the graph of $y = \frac{1}{x}$ is symmetric with respect to the origin (a 180-degree turn gives you the same graph), and the graph of $y = \frac{1}{x^2}$ is symmetric with respect to the y-axis (it's a mirror image on either side).

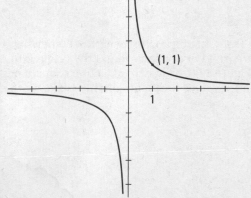

Figure 17-5:
The graph of $y = \frac{1}{x}$ is in only the first and third quadrants.

(1, 1)

1

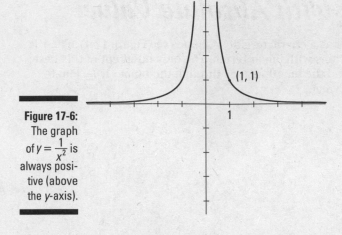

Figure 17-6:
The graph
of $y = \frac{1}{x^2}$ is
always positive (above
the y-axis).

(1, 1)

1

Rooting Out Square Root and Cube Root

The graph of $y = \sqrt{x}$ starts at the origin and stays in the first quadrant (see Figure 17-7). Except for $(0, 0)$, all the points have positive x- and y-coordinates. The curve rises gently from left to right.

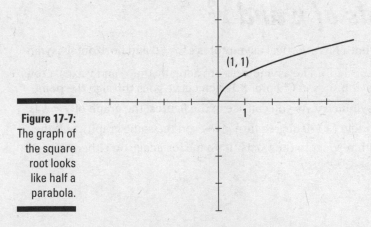

Figure 17-7:
The graph of
the square
root looks
like half a
parabola.

(1, 1)

1

The graph of $y = \sqrt[3]{x}$ is an odd function (see Chapter 5): It resembles, somewhat, twice its partner, the square root, with the square root curve spun around the origin into the third quadrant and made a bit steeper (see Figure 17-8). You can take cube roots of negative numbers, so you can find negative x- and y- values for points on this curve. Both curves go through the point $(1, 1)$.

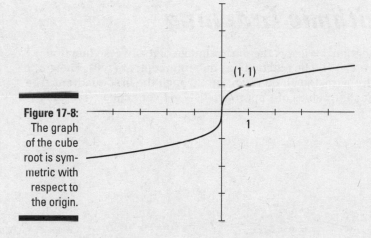

Figure 17-8:
The graph
of the cube
root is sym-
metric with
respect to
the origin.

Growing Exponentially with a Graph

The graph of the exponential function $y = e^x$ is always above the x-axis (see Figure 17-9). The only intercept of this graph is the y-intercept at $(0, 1)$. The x-axis is the horizontal asymptote when x is very small, and the curve grows without bound as the x-values move to the right. (Check out Chapter 9 for practice with the number e.)

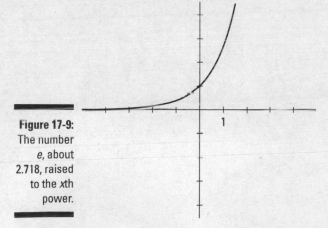

Figure 17-9:
The number
e, about
2.718, raised
to the xth
power.

Logging In on Logarithmic Graphing

The graph of the logarithmic function $y = \ln x$ is the mirror image of its inverse function, $y = e^x$, over the line $y = x$ (see Figure 17-10). The function has one intercept, at $(1, 0)$. The graph rises from left to right, moving from the fourth quadrant up through the first quadrant. The y-axis is the vertical asymptote as the values of x approach 0 — get very small. (For the scoop on logs, flip to Chapter 9.)

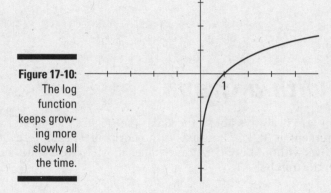

Figure 17-10: The log function keeps growing more slowly all the time.

Chapter 18

Ten Special Sequences and Their Sums

● ●

In This Chapter

▶ Sequences involving polynomials and formulas for their sums

▶ Describing special patterns and adding them

▶ Sequences for *e* and sin *x*

● ●

A *sequence* is a list of terms that has a formula or pattern for determining the numbers to come. A *series* is the sum of the terms in a sequence. Chapter 15 explains several types of sequences and series. In this chapter, I give you the power to add up lists of numbers — sometimes a huge number of numbers. Many sequences of numbers are used in financial and scientific formulas, and being able to add them up is essential. This chapter doesn't contain all the sum formulas possible, but the formulas here are the most common.

In Chapter 17, I tell you about basic graphs you can change through translations. The formulas in this chapter can be changed, too, but you'll have to wait until you get to a more advanced math course to get the opportunity. For now, just enjoy.

Adding As Easy As One, Two, Three

The positive integers are 1, 2, 3, 4, 5, . . . , going on forever. Find the sum of the first *n* terms of this sequence using this formula:

$$1+2+3+\cdots+n=\frac{n(n+1)}{2}$$

So, if you want the sum of the first 100 integers, you do the following calculation:

$$1+2+3+\cdots+100=\frac{100(100+1)}{2}=\frac{\overset{50}{\cancel{100}}(101)}{\cancel{2}}=50(101)=5{,}050$$

Summing Up the Squares

The squares of the positive integers are 1, 4, 9, 16, 25, . . . , n^2. To find a particular term in this sequence, you just take the square of the number of the term (the 12th term is $12^2 = 144$).

Find the sum of the first n terms using $1^2 + 2^2 + 3^2 + \cdots + n^2 = \dfrac{n(n+1)(2n+1)}{6}$. To add up the numbers $1 + 4 + 9 + 16 + 25 + \cdots + 144$, you want the sum of the first 12 squares. Using the formula,

$$1 + 4 + 9 + 16 + \cdots + 144 = 1^2 + 2^2 + 3^2 + \cdots + 12^2$$
$$= \frac{12(12+1)(2(12)+1)}{6} = \frac{12(13)(25)}{6} = \frac{\overset{2}{\cancel{12}}(13)(25)}{\cancel{6}} = 2(13)(25) = 650$$

Finding the Sum of the Cubes

The cubes of the positive integers are 1, 8, 27, 64, 125, . . . , n^3. The rule for the general term is n^3; you just raise the number of the term to the third power. You can find the sum of these cubes, 1^3, 2^3, 3^3, and so on, using $1^3 + 2^3 + 3^3 + \cdots + n^3 = \dfrac{n^2(n+1)^2}{4}$. If you look closely, you can see that this formula is just the square of the formula used to find the sum of the first n integers — each factor in the formula is squared.

If you want to find the sum of $1 + 8 + 27 + 64 + 125 + 216 + \cdots + 3{,}375$, you first have to figure out whose cube 3,375 is. A simple calculator can find this answer for you. Or, with some sleuthing, you may determine that the root has to end in a 5 for the cube to end in a 5. Seeing that $3{,}375 = 15^3$ doesn't take long. Here's how you find the sum:

$$1^3 + 2^3 + 3^3 + \cdots + 15^3 = \frac{15^2(15+1)^2}{4}$$
$$= \frac{225(256)}{4} = \frac{225(\overset{64}{\cancel{256}})}{\cancel{4}} = 225(64) = 14{,}400$$

Not Being at Odds with Summing Odd Numbers

The odd positive integers are 1, 3, 5, 7, 9, . . . , $(2n - 1)$. If you want the fifth odd number, the 9, you replace the n with 5 in $2n - 1$ and get $10 - 1 = 9$.

When the numbers get large, the process for finding out where a particular odd number falls in the sequence becomes even more important — you really don't want to have to add up all the numbers for the series, but you have to know how many terms are in the list. You can calculate, for instance, that the number 2,357 is the 1,179th odd number. You find this answer by solving $2n - 1 = 2{,}357$ and $2n = 2{,}358$. Dividing by 2, you get $n = 1{,}179$, so 2,357 is the 1,179th odd number.

The formula for the sum of n odd numbers is $1 + 3 + 5 + \cdots + (2n - 1) = n^2$. To add up the odd numbers $1 + 3 + 5 + 7 + \cdots + 2{,}357$, you first determine how many numbers are in the list: $2n - 1 = 2{,}357$, so $n = 1{,}179$. The sum is $1{,}179^2 = 1{,}390{,}041$.

Evening Things Out by Adding Up Even Numbers

The even positive integers are 2, 4, 6, 8, . . . , $2n$. Determine the sum of the first n even numbers with the formula $2 + 4 + 6 + 8 + \cdots + 2n = n(n + 1)$. You may notice that this formula looks somewhat like the formula for adding up the first n integers (see "Adding As Easy As One, Two, Three," earlier in this chapter). Multiply that other formula by 2, and you get the formula for the evens. Seems to make sense, even. (Ouch — horrible pun.)

Adding up the numbers $2 + 4 + 6 + 8 + \cdots + 500$, you determine that $n = 250$. So the sum is $500(501) = 250,500$.

Adding Everything Arithmetic

Arithmetic sequences are very predictable. The terms are always a constant difference from one another. The terms in an arithmetic sequence are $a_1, a_2, a_3, \ldots, a_n = a_1, a_1 + d, a_1 + 2d, a_1 + 3d, \ldots, a_1 + (n - 1)d$. The number of times you add the common difference to the first term is 1 less than the total number of terms.

To find the sum of the terms of an arithmetic sequence, you can use one of these formulas:

$$a_1 + (a_1 + d) + (a_1 + 2d) + \cdots + (a_1 + (n-1)d) = \frac{n}{2}\left[2a_1 + (n-1)d\right]$$

$$a_1 + (a_1 + d) + (a_1 + 2d) + \cdots + (a_1 + (n-1)d) = \frac{n}{2}\left[a_1 + a_n\right]$$

Finding the sum using either formula requires that you know how many terms are in the sequence. Solve for n, the number of terms, using the rule for the general term of the arithmetic sequence: $a_n = a_1 + (n - 1)d$.

To find the sum of $4 + 10 + 16 + 22 + 28 + \cdots + 304$, you recognize that the first term, a_1, is 4, the last term, a_n, is 304, and the difference between each pair of terms is 6. Solving $4 + (n - 1)6 = 304$, you get $6n - 6 = 300$; $6n = 306$; $n = 51$. So, the sum of those 51 terms is

$$4 + 10 + 16 + \cdots + 304 = \frac{51}{2}\left[4 + 304\right] = \frac{51[308]}{2} = \frac{51[\cancel{308}^{154}]}{\cancel{2}} = 7,854$$

Geometrically Speaking

Geometric sequences are almost as predictable as arithmetic sequences. The terms have a common ratio — you divide any term in a geometric sequence by the term that comes before it, and you get the same ratio or multiplier, no matter which two terms you choose. The geometric sequence 1, 3, 9, 27, 81, . . . has a first term of 1 and a common ratio of 3. The general term for a geometric sequence is $g_n = g_1(r^{n-1})$, where g_1 is the first term and r is the common ratio.

To add up the first n terms of a geometric sequence, use this formula:

$$g_1 + g_1 r + g_1 r^2 + \cdots + g_1 r^{n-1} = \frac{g_1(1 - r^n)}{1 - r}$$

Adding up $1 + 3 + 9 + 27 + \cdots + 6,561$, you first determine that 6,561 is 3^8. This answer makes 6,561 the ninth term in the series, because $1(3^8) = 1(3^{n-1})$, and if $8 = n - 1$, then $n = 9$. Using the formula for the sum,

$$1 + 3 + 9 + \cdots + 1(3^{9-1}) = \frac{1(1 - 3^9)}{1 - 3}$$

$$= \frac{1(1 - 19683)}{-2} = \frac{-19682}{-2} = 9,841$$

Easing into a Sum for e

The number e is approximately equal to 2.718281828459. The value for e is irrational, which means that the decimal never ends or repeats. The number, however, is oh-so-necessary in the scientific, mathematical, and financial worlds (see Chapter 9 for an application that involves your bank account).

You can determine the value for e using various formulas, one of which is the sum of a sequence of numbers. The more terms you add up in this series, the more correct decimal places for e you determine. The general term for this series is

$$e \approx \frac{1}{0!} + \frac{1}{1!} + \frac{1}{2!} + \frac{1}{3!} + \cdots + \frac{1}{(n-1)!}$$

I don't have a handy formula for adding up the terms in this series, but a scientific calculator can do the job for a while. Here are a few of the sums:

$$\frac{1}{0!} + \frac{1}{1!} = \frac{1}{1} + \frac{1}{1} = 2$$

$$\frac{1}{0!} + \frac{1}{1!} + \frac{1}{2!} = \frac{1}{1} + \frac{1}{1} + \frac{1}{2} = 2.5$$

$$\frac{1}{0!} + \frac{1}{1!} + \frac{1}{2!} + \frac{1}{3!} = \frac{1}{1} + \frac{1}{1} + \frac{1}{2} + \frac{1}{6} \approx 2.666666667$$

$$\frac{1}{0!} + \frac{1}{1!} + \frac{1}{2!} + \frac{1}{3!} + \frac{1}{4!} = \frac{1}{1} + \frac{1}{1} + \frac{1}{2} + \frac{1}{6} + \frac{1}{24} \approx 2.708333333$$

$$\vdots$$

$$\frac{1}{0!} + \frac{1}{1!} + \frac{1}{2!} + \frac{1}{3!} + \cdots + \frac{1}{10!} = \frac{1}{1} + \frac{1}{1} + \frac{1}{2} + \frac{1}{6} + \cdots + \frac{1}{3628800} \approx 2.718281801$$

Signing In on the Sine

The *sine* is one of the trigonometric functions. The input (x-value) for trig functions consists of angle measures. For instance, if $y = \sin x$ (sin is the abbreviation for sine) and x is 30 degrees, then $y = \sin 30° = 0.5$.

Several different methods are available for computing the value of the sine of an angle, but the one that fits in this chapter is an infinite series of values. If you write the input, x, in radians instead of degrees ($360° = 2\pi$ radians), then $\sin x \approx x - \dfrac{x^3}{3!} + \dfrac{x^5}{5!} - \dfrac{x^7}{7!} + \cdots + (-1)^{n-1}\dfrac{x^{2n-1}}{(2n-1)!}$. The more terms you use, the closer you get to the exact value of the function.

So, to find the sine of 30 degrees, change that measure to radians to get $30° \approx 0.5236$ radians (30 degrees is one-twelfth of a circle, and $2\pi \div 12 = \dfrac{\pi}{6}$ radians). Using the series,

$$\sin 0.5236 \approx 0.5236 - \frac{0.5236^3}{3!} + \frac{0.5236^5}{5!} - \frac{0.5236^7}{7!} + \cdots$$

$$\approx 0.5236 - \frac{0.1435}{6} + \frac{0.0394}{120} - \frac{0.0108}{5040}$$

$$\approx 0.5236 - 0.0239 + .0003 - 0$$

$$= .5$$

In the fourth term, the number is so small that it rounds to 0 when you just want four decimal places.

Powering Up on Powers of 2

You can quickly and simply find the sum of a sequence consisting of powers of 2 just by finding the next power of 2 and subtracting 1. For instance, $1 + 2 + 4 + 8 + 16$ consists of the powers of 2 from 2^0 to 2^4. The sum is $2^5 - 1 = 32 - 1 = 31$. So, the rule is $1 + 2 + 4 + 8 + \cdots + 2^n = 2^{n+1} - 1$.

Adding Up Fractions with Multiples for Denominators

Create a rather interesting sequence with

$$\frac{1}{1 \cdot 2} + \frac{1}{2 \cdot 3} + \frac{1}{3 \cdot 4} + \cdots + \frac{1}{n \cdot (n+1)}$$

where you take the number of the term times the next integer and put their product in the denominator of a fraction. The sum of the terms is equal to $\dfrac{n}{n+1}$. That result almost seems too easy to believe, doesn't it? Look at a few examples:

$$\frac{1}{1 \cdot 2} + \frac{1}{2 \cdot 3} + \frac{1}{3 \cdot 4} + \cdots + \frac{1}{n \cdot (n+1)} = \frac{n}{n+1}$$

$$\frac{1}{1 \cdot 2} + \frac{1}{2 \cdot 3} = \frac{1}{2} + \frac{1}{6} = \frac{3}{6} + \frac{1}{6} = \frac{4}{6} = \frac{2}{3}$$

$$\frac{1}{1 \cdot 2} + \frac{1}{2 \cdot 3} + \frac{1}{3 \cdot 4} = \frac{1}{2} + \frac{1}{6} + \frac{1}{12} = \frac{6}{12} + \frac{2}{12} + \frac{1}{12} = \frac{9}{12} = \frac{3}{4}$$

$$\frac{1}{1 \cdot 2} + \frac{1}{2 \cdot 3} + \frac{1}{3 \cdot 4} + \frac{1}{4 \cdot 5} = \frac{1}{2} + \frac{1}{6} + \frac{1}{12} + \frac{1}{20} = \frac{30}{60} + \frac{10}{60} + \frac{5}{60} + \frac{3}{60} = \frac{48}{60} = \frac{4}{5}$$

Index

• •

• D •

About the Author

Mary Jane Sterling is also the author of *Algebra I For Dummies, Algebra II For Dummies, Math Word Problems For Dummies, Business Math For Dummies,* and *Linear Algebra For Dummies.* She taught junior high school and high school math for several years before beginning her current 30-plus-year tenure at Bradley University in Peoria, Illinois. Mary Jane especially enjoys working with future teachers and trying out new technology. She and her husband, Ted, enjoy spending their leisure time with their children and grand-children, traveling, and fishing.

Dedication

I'd like to dedicate *Algebra II Workbook For Dummies* to my children, Howard Jonathan, James Theodore, and Jane Christine (yes, three *J*'s: Jon, Jim and Jane). My husband and I raised our children to be "independent," and, oh my, we got what we wished for in that quest. We're so proud of the people our children have become.

Author's Acknowledgments

I warmly thank my project editor, Elizabeth Kuball, for working on this newest endeavor; she is so thorough and professional, yet understanding and patient. Thank you to the technical editor, Michael McAsey; without his skills and careful checking, mathematical mistakes may have gone unnoticed and uncorrected. And thank you to Lindsay Lefevere for her continued efforts to keep this author happily occupied.

Publisher's Acknowledgments

Executive Editor: Lindsay Sandman Lefevere

Project Editor: Elizabeth Kuball

Copy Editor: Elizabeth Kuball

Technical Editor: Michael McAsey, PhD

Project Coordinator: Sheree Montgomery

Cover Image: ©iStockphoto.com/3dbobber